Angelo Heilprin

The Bermuda Islands

A Contribution to the Physical History and Zoology of the Somers Archipelago...

Angelo Heilprin

The Bermuda Islands
A Contribution to the Physical History and Zoology of the Somers Archipelago...

ISBN/EAN: 9783744716741

Printed in Europe, USA, Canada, Australia, Japan

Cover: Foto ©berggeist007 / pixelio.de

More available books at **www.hansebooks.com**

THE NORTH ROCK.

THE
BERMUDA ISLANDS:

A CONTRIBUTION TO THE PHYSICAL HISTORY AND ZOOLOGY

OF THE

SOMERS ARCHIPELAGO.

WITH AN EXAMINATION OF THE STRUCTURE OF CORAL REEFS.

RESEARCHES UNDERTAKEN UNDER THE AUSPICES OF THE ACADEMY OF
NATURAL SCIENCES OF PHILADELPHIA.

BY

ANGELO HEILPRIN, F.G.S.A., F.A.GEOGR. SOC.,

Curator-in-Charge of, and Professor of Invertebrate Paleontology at, the Academy of
Natural Sciences of Philadelphia; Professor of Geology at the Wagner Free
Institute of Science; Member of the American Philosophical Society, etc.

WITH ADDITIONS BY

Prof. J. Playfair McMurrich, Mr. H. A. Pilsbry, Dr. George
Marx, Dr. P. R. Uhler, and Mr. C. H. Bollman.

PHILADELPHIA:
PUBLISHED BY THE AUTHOR.
ACADEMY OF NATURAL SCIENCES.
1889.

THE observations recorded in the following pages are the outcome of a vacation-journey undertaken in the summer of 1888 in company with a class of students from the Academy of Natural Sciences of Philadelphia. My main object in visiting the islands was to satisfy my mind on certain points connected with the structure and physiognomy of coral reefs, to the study of which the Bermudas offer special advantages. I contemplated but little work in zoology, and that which was accomplished may be considered supplemental to the plan of work originally laid out. It was not until my return to Philadelphia that I was made aware of the extent of the zoological material collected by us, and how little systematic study of the fauna of the region had been made prior to our visit. Some of our material still awaits examination; for the elaboration of that portion which is delineated in the present volume I am largely indebted to the labors of a number of specialists who have kindly volunteered their assistance. In this connection I desire to acknowledge my indebtedness to Prof. J. Playfair

McMurrich, formerly of Haverford College; to Mr. H. A. Pilsbry, of the Academy of Natural Sciences; to Dr. George Marx, of Washington; to Dr. P. R. Uhler, of Baltimore; to the late Mr. C. H. Bollman, of Bloomington, Ind.; and to two of my immediate assistants, Messrs. J. E. Ives and Witmer Stone.

To Miss A. Peniston, of Peniston's, Bermuda, I am under special obligation for the use of much new material illustrating the Bermudian fauna; and I am similarly indebted to Dr. W. H. Dall, of the U. S. National Museum, Washington, for having placed at my disposition the collection of Bermuda shells made a few years ago by Mr. G. Brown Goode.

Finally, I desire to convey my thanks to the members of my class—Messrs. J. E. Ives, Witmer Stone, Roberts Le Boutillier; Misses Emma Walter, Mary A. Schively, Virginia Maitland, Emily G. Hunt, Ella Hunt—all of whom rendered much valuable assistance either in the field or in the laboratory.

A. H.

ACADEMY OF NATURAL SCIENCES,
Philadelphia, September, 1889.

CONTENTS.

CORRECTIONS.

PAGE.

40. For "perspicuity", sixth line from top, rea l *perspicacity*.

58. In place of the words "more rapid", third line from top, substitute *greater*.

102. For "pediculate", eighth line from bottom, read *pedicellate*.

THE BERMUDA ISLANDS.

My first impression of the Bermudas was one of disappointment. I had heard so much of the "hundred islands," of the luxuriant vegetation, that I found it difficult to realize that these undulating hills, rising in their garb of withered green, were in reality the far-famed pearl of the Atlantic. But our visit was timed for the month of July, and possibly the withered condition of the vegetation had something to do with this feeling of disappointment. The clumps of palmettos which are bunched against the hillsides were as yet undistinguishable, and the eye rested on a monotonous expanse of dirty green, relieved here and there by dark masses of the Bermuda juniper, which, from a distance, recalled the cloud-shadowed patches of our northern mountain slopes. Innumerable particles of white cottages gleamed forth in the bright sunlight, but their uniform brilliancy only served to intensify the sombreness of the background which they illumined.

I had, from assumed geological knowledge, expected to see a long white crest rolling over the outer reef, but in this expectation I was also disappointed. We were being carried in on the flood, and no trace of this natural parting of the waters was visible. As we approached nearer to the shore, however, abrupt changes in the color of the water revealed the position of the coral-shallows, but we as yet saw nothing of the reef of the imagination. Occasional jelly-fishes floated lazily by, and the ever merry petrels were still quivering in our path of beaten foam. Our first tropic-bird hovered about the rigging, seemingly surprised at the early intrusion which we had permitted ourselves. The flying-fishes became more numerous as we neared the islands, and they could be frequently seen skipping away five or more at a time, and usually in a direction at right angles to the line of the ship.

I was anxious to determine the true nature of their aerial locomotion, and to settle the vexed question of the supposed flying movements of the fins. We accordingly watched these interesting creatures very intently, and followed them with our glasses over their entire course. So rapid was their motion, however, that it was not easy to keep them within the field of the glass, and still less easy to hold them in distinct vision, and for a long time we really hardly knew what we saw. We failed, however, to detect any positive continuous movement on the part of the fins, and it certainly appeared as though in all, or nearly all, cases the animal merely shot forward as the result of some primary impetus, taking a course nearly horizontal with the surface of the water. This direct course, in view of the apparent method of propulsion, was certainly surprising, as it would naturally be expected that from an initial leap the line of travel would be that of a regular curve. But the horizontal course was possibly more apparent than real. At intervals, two or three times in a single flight, the animal seemed to strike the crest of a wave, and acquire new impetus from a blow of the tail. In this way the flight may be said to be one of distinct passages —although without arrest of movement—in which the curve motion largely disappears, or is at least reduced to its lowest terms. Indeed, the successional character of the flight could be plainly seen in the sudden jerky changes of direction which were marked off at irregular intervals, and at points where rising waves apparently met the supplemental strokes of the tail. The duration of flight was from five to eleven or twelve seconds, while the distance covered was probably, in extreme cases, not less than 200–300 feet, and possibly even more.

The inner waters of the reef showed those remarkable contrasts of color which have been so frequently dwelt upon and depicted by travelers. From the most intense indigo we pass abruptly to a brilliant emerald, and from this again possibly to a bright sea-green. So sudden are the transitions that the semblance of a natural water is largely destroyed, and for a moment one feels inclined to doubt the reality of the scene before him. I must confess that had I been informed in advance of these wonderful chromatic effects, I should have been loud in pronouncing the impossibility of their occurrence, so wholly unnatural did they appear.

Passing through the line of old hulks, whose grass-grown bottoms and battered planks emphasized the words of condemnation which relegated the ancient merchantmen to the rank of *objets d' art*, we entered the harbor of Hamilton. The fact of its being Sunday did not obtrude itself upon the throng that had assembled to greet us on our arrival. The helmeted redcoat and servant, custom-house officials and steamship agents were out in force, but they were far outnumbered by that class of easy-going inhabitants whose hardest labor appears to be that of doing nothing. The time-honored custom of building connecting gang-planks instead of bringing the ship close up to the wharf, delayed our debarkation by about a half-hour, but delays of this or a similar kind, as we soon discovered, are of little moment with the Bermudians.

The capital city, Hamilton, has little of interest to detain the stranger beyond the beautiful display of exotic plants which are to be found in the private gardens. The broad and pleasant avenues which intersect the town at nearly right angles, and glisten with that intensity of which only a white limestone is capable, possess the general features of the ordinary

country roads, setting firm and hard under almost all conditions of weather and temperature. They rarely require repair-
ing, and their even compactness well sustains the quality for
which the Bermudian roads are famous. Rain affects them but
little, except in so far as it assists as a solidifier, and a few
moments after a heavy shower they are generally as pass-
able as after prolonged exposure to the sun. Here and there
extensive villas and gardens betray opulence, but more commonly it is the appearance of pleasurable comfort rather than
the possession of riches which appeals to the eye of the visitor.
The recessed house-fronts, opening upon broad, shaded veran-
das, typify a style of architecture doubtless best adapted to
the requirements of the climate, while the dazzling whiteness
of the exteriors emphasizes an amount of attention bestowed
upon house-decoration which would probably surprise even
the proverbially neat cottagers of rural Belgium and Holland.
We were informed that the operation of whitewashing was re-
sorted to as often as twice a year, and where, as is almost uni-
versally the case throughout the island group, the drinking
water is collected as off-flows from the roof, this part of the house
is anointed as well as the sides. At intervals throughout the
town, as also in the country, extensive inclined basins have
been constructed for the reception of rain-water, and serve
as reservoirs for periods of emergency.

The shops are in the main not attractive, and on the whole
they betray a lack of energy in their management which is
surprising in a region so largely visited by strangers. We
were recommended to a presumably fashionable confectioner's,
the floor-space in whose establishment was given up in large
part to a display of hats and clothing, and the walls to musical
instruments. In another large establishment it was reported
that anything could be obtained from a coffin to a pulpit, but
we found it impossible to procure an ordinary insect net; nor
were we better rewarded as far as this, to us necessary, article
was concerned, by inquiry elsewhere.

As before remarked, the attractive feature of the town is to be found in the display of exotic plants. This exhibit is, however, not confined to Hamilton; it is the property of the entire island group, where it has become, if the expression might be permitted, naturalized. The unfortunate ones among us who in the northern region ecstatically wonder at the rare treats which the floriculturist from time to time lays before them, can obtain but little comfort from a trip to the Bermudas. Their earlier-formed notions of grandeur soon disappear. It would, however, be conveying a false impression to state that the vegetation is luxurious, or that it is at all comparable in exuberance with the vegetation of the true tropics, or even with that of many parts of the peninsula of Florida. On the contrary, it is on the whole sparse, and only here and there, in favored localities, or where the husbandman has largely assisted nature, does it reveal those touches of picturesque quality which so impress the mind of the stranger, and lead him to believe in special luxuriance.

The native arboreal vegetation is scrubby, consisting almost wholly of the Bermuda juniper and the sabal or palmetto, the latter being probably the only native species of palm of the twelve or more forms now found on the islands. The date and cocoa-nut are both cultivated, but it is only exceptionally that the fruit arrives at maturity. Superb specimens of the former, the so-called "three sisters," are found in the singularly attractive public garden of St. George's, but elsewhere the tree is not exactly uncommon, rising generally in solitary grandeur above its less pretentious associates. Nothing, it appears to me, can surpass in majesty the five specimens of cabbage-palm (*Oreodoxa oleracea*) which adorn the roadside at Pembroke Hall, in the outskirts of Hamilton, and justly constitute the pride of the Bermudians. Like granite monoliths the gray shafts, 60–70 feet in height, stand unbending to the elements that play about them, unmoved by the force that tosses their delicate plumes into restless activity.

Of our common deciduous trees, oaks, birches, beeches, poplars, etc., there is scarcely a trace to be seen, nor is there any-

thing to replace them. An occasional sycamore, mulberry or
weeping-willow still reminds one of the temperate north, but
beyond these and the Bermuda juniper, there is little suggestive of the woods barely six hundred miles distant. The uni-
versality of the juniper, however, imparts a decidedly northern aspect to the vegetation, despite the large number of tropical elements that are embodied in it. The latter are too
irregularly scattered to constitute dominating factors in the
modeling of the landscape. At intervals bits of delicious tropical nature surprise one; warm and sunshiny patches of palm-
land, largely overgrown with the coarse bracken, and bordered
by almost impenetrable thickets of banana and plantain, fall
refreshingly upon the eye that has perhaps become satiated
with the juniper, and that glory of the Bermudas, the oleander.
O'er hill and dale, far and near, in the garden and along the
roadside, this superb bush scatters its fragrance to the winds.
For one who has not seen the rose-flowered oleander in its native home, or in this land of its adoption, it is impossible to
conceive of the effect which is produced by the great masses of
showy blossoms which appear here, there, and everywhere
throughout the landscape. The hedges are ablaze with their
blossoms, and buried in perfume; the roadways are similarly bejeweled and scented. I can only compare the general
effect with that produced by our copses of rhododendron, but,
while the flowers and foliage of the oleander suffer as individual elements in the comparison, they more than compensate
by their masses.

A first impression of a country drive in the Bermudas,
along some such road as the "middle road" leading from
Hamilton to Flatts Village, is not soon forgotten. The gray,
one might almost say omnipresent, bounding wall, which
forms part of the natural rock of the islands, the retired and
sheltered cottages, the peculiar inhabitants—more black than
white—and above all the vegetation, strike one as strange and
novel. Birds of bright plumage, the red cardinal and blue-bird,
flit numerously before you, and although they are regular

members of the North American avifauna as well, they appear here in a different light than with us, since they, with the little ground-dove, form the most noticeable, if not the dominating, elements in the bird-fauna of the islands. We are in a little world of our own, distinct from either Europe or America.

The corn-fields of our temperate north are here largely replaced by fields of bananas and plantains, which are grown in dense and almost impenetrable thickets. Although at first attractive by their novel, and suggestively tropical appearance, the eye soon wearies of the gross and wind-rifted leaves, and eagerly falls upon the much more delicate vegetation of the bamboo, which here and there shoots its willowy tufts 30 or 40 feet into the air. The cane is also a graceful ornament about many of the country cottages.

The banana "trees" were laden with fruit at the time of our visit, and naturally we had a good opportunity to judge of the merits of this highly-prized article of food with the Bermudians. At breakfast, dinner, and supper it was a constant accompaniment of the table, and always welcome. The fruit is considerably smaller than the imported article with us, and decidedly tastier or sweeter, doubtless due to its being picked fresh from the plant. The more commonly cultivated variety is the form known as *Musa Cavendishi*, or dwarf banana, which rarely exceeds, to the bend or apex of the leaf, more than about eight feet in height; it always presents a stubby appearance, more like a great coarse weed, which has grown up spontaneously, than a cultivated plant. Indeed, it can scarcely be said to be under cultivation, since practically no attention is paid to its proper growth and development; nature does it all. Bunches of fruit weighing fifty pounds or more are no rarity, and occasionally they reach nearly double this weight.

The plantain, which is more rarely seen, and whose fruit is not held in popular favor as an article of food, is a much more graceful and imposing plant, with leaves eight or ten feet, or even more, in length. It grows to more than twice the height of the common banana, and, although its leaves are largely wind-rifted, it never presents the shabby appearance of the latter.

Among the other distinctive accompaniments of the roadside are the aloe and yucca, or Spanish bayonet. The flowering stems of the former, rising twenty or twenty-five feet in height, are not an uncommon feature in the landscape of the garden-tracts. Of the fifteen or more species of cactus now found on the islands, some of which, like the Turk's cap, or melon cactus, and the night-blooming cereus, are extensively cultivated in the gardens, only one, the prickly-pear (*Opuntia vulgaris*), can be said to be at all common. It is found scattered here and there about the sandy wastes, or hanging in sprawling masses from the road walls. Some notion of the luxuriance of growth may be had from the condition of the plants of the night-blooming cereus. Shortly before our departure from the islands we were hospitably invited to visit a plant of this species in the garden of a Mrs. Peniston, just outside of Flatts Village. It is scarcely necessary to say that we were surprised when we beheld a plant covering an area equal to that of a fair-sized room, and supporting upwards of 200 of the most superb flowers! Well may the tourist go back down-hearted to his native conservatory.

But this is not the only instance where a comparison between home and foreign products may give rise to sad reflections. Alas, how sadly changed must be the conception of a rubber-plant, after one, who had hitherto been accustomed only to the "wonders" of the drawing-room and conservatory, has seen the monster, stretching forth its arms like an oak, at "Par le Ville," Hamilton!

If, however, the eye is riveted to these remarkable forms of vegetality, it meets only with disappointment when it scans the usually sparse herbaceous vegetation of the woodlands, or that which covers the open meadows and rock-surfaces. We look in vain for that rich, dense green which forms the sunny undergrowth of our forests, for the crop of weed and grass which is the wealth of our fields and pastures. Plenty of grasses there are, it is true, but they are in most places thin and weak. Only here and there in the deep and open hollows, as in the neighborhood of the "marshes," do they show anything like luxuriance of growth It should, however, be stated that in many sheltered limestone hollows, as well as along similarly protected rock-ledges, the vegetation is wonderfully profuse, interwoven and intermatted so as to render penetration almost impossible. The coffee-tree thicket back of Joyce's cave, and the convolvulus cave close by, are wonderful exhibitions of this kind; and likewise the tract about Walsingham. But over the greater part of the island one may sing his pæans to the sage (Lantana), which has virtually taken possession of the soil, a not very attractive nor delightfully odorous plant.

Our headquarters during our brief stay at the Bermudas were at the Flatts Village, a small assortment of neat houses and cottages situated at the entrance to Harrington Sound. A strong current passes here at high-water into the inner basin, bringing in much sediment from the open sea, and with it a rich assortment of animal life. The low-water sands constitute one of the best collecting grounds of the archipelago, excepting, of course, the outer reefs, and the under surface of almost every stone offered something for our collections. We were fortunate to discover here a new species of cuttle-fish (*Octopus chromatus*) characterized by unusually attenuated tentacles; but, beyond two other individuals of another species which we failed to obtain, this was the only representative of this order of animals with which we came in contact during the entire journey. Yet, they are said to be specially abundant just among the rock crevices of Flatts Inlet.

It is wonderful with what tenacity these animals hold on to their anchorages when a forcible attempt is made to eject them from their shelters. We in vain tried by both coaxing and pulling to dislodge one of these interesting creatures, but, despite the havoc made by chopping the enveloping rock, we

were compelled to desist from our attempt after the labor of a
full hour and a half. The chromatic changes of the skin were
at times very rapid, and so nicely did the animal adjust its
color to that of the rock as to render its outline barely visible.

The wonderful transparency of the greenish waters permitted
objects to be distinctly visible at depths of 20–25 feet, and where
there were no moving ripples, at even greater depths. I
doubt much, however, the statement that under ordinarily
favorable conditions objects may be distinguished at depths of
60 to 70 feet, even with the aid of the water-glass ; at any rate,
our own experience failed to give support to this commonly-
received notion. Almost everywhere within the bounding
reef, except where special circumstances have favored the development of coral and millepore patches, the bottom is largely
barren, presenting a nearly uniform expanse of coral sand.
This is the paradise of the sea-urchin (*Toxopneustes variegatus*)
and of the great black sea-cucumber which is so common in
the inner waters. From the anchorage of the Villa Frascati we
watched day after day the uncouth ebony masses of the latter,
scattered like so many black blotches over the sand. Only exceptionally could they be observed to change their position,
although the rolls of sand drawn over the surface indicated
that such changes were not uncommon ; not improbably their
perambulations take place largely at night-time, when their
movements would not be likely to attract attention. In the
normal condition of rest and apparent non-animality the creature might readily pass aggressive observation, even though
it be the most conspicuous object on the bottom ; and in thus
deceiving what might otherwise be formidable enemies it is
probably largely assisted by its forbidding black color. It is
a significant fact, although it may hold no special relation in
the matter, that another form of holothurian (like the latter,
new to science), of almost equal size, and with a ground color
of reddish-yellow, occupies the same haunts, but in vastly di-
minished numbers. Can it be that it is suffering extermination through the possession of a more attractive, even though
less apparent, coloration ?

Some of the pleasantest of our leisure hours in the Bermudas were passed in gazing into the kaleidoscopic waters which bathe the gardens of the Villa Frascati, and who that has once glanced into this liquid glass can forget the picture that is there presented? A giant palette dipped beneath the water. We have the same bright smudges of cardinal and lemon-yellow, the streaks of green and blue, the purples, oranges, and blacks —in short, all that brilliancy and wealth of color which belong to the painter's upper row. A species of encrusting sponge, possibly a Microscionia, made large patches of the brightest red on some of the detached rock, while immediately alongside, another sponge formed clumps of equally brilliant yellow, and still another, lesser patches of green. This violent contrast of color, which is still further intensified by the whiteness of the supporting coral sand, is kept in a pure key. There is no discord, and not very much more true harmony; it is strongly marked individuality. If any one still doubted that ocean water had a natural color of its own, a single glance at the flowing emerald would have been sufficient to dispel all doubts in the matter.

The most beautiful of the inner waters of the archipelago is Harrington Sound, an almost closed lagoon which extends for about three miles eastward from Flatts Village. Its only open communication with the sea is by way of Flatts Inlet, the extreme contraction of which compels the water to flow in and out with the fury of a mill-race. We did much of our dredging here, but scarcely met with that success which we had anticipated. The bottom is manifestly largely a sand-barren, and in a general sense unproductive. We, however, picked up specimens of a beautiful new species of Chromodoris, and quantities of a remarkably crassiform Chama. The latter more nearly approaches a fossil from the Pliocene deposits of Florida than any other species with which I am acquainted. Along the borders of the Sound the coral-growth, consisting mainly of Isophyllia, Oculina and Siderastræa, is largely developed, but we everywhere failed to detect traces of the

large meandriniform types which form such a prominent
feature in the life of Castle Harbor. In the latter, again, the
Isophyllias were largely wanting. Probably upon the much
greater depth of water in Harrington Sound, and the absence
of a submarine platform, is dependent mainly the difference
in the coral life of the two almost contiguous bodies of water.

It is difficult to conceive of a more beautiful and refreshing
sight than that which is presented by the sheltered coral pools,
with their wealth of color, which bite here and there beneath
the crumbling banks. Corals, millepores, and stone-encrusting
Algæ vie with each other in building up the common domain,
whose most intricate details are revealed by the transparency
of the waters. Even the tiny polyps may be seen expanding
their delicate crowns of tentacles. What a charming vision of
the busy life about us! The sea-anemones are especially
abundant in these rock-shelters, and some of them are strikingly beautiful; but on the whole they are less brilliantly
colored than those of our northern shelters. Among the corals,
some of the species of Isophyllia were the most highly colored,
the browns and greens being remarkably rich. Below a depth
of a comparatively few feet coral-life largely disappeared, and
along the actual floor of the Sound, say at depths of from six
to eight fathoms, we only obtained Oculina. No haul of over
ten fathoms ever brought up a coral.

During much of our stay on the Sound the water was very
rough, and its inroads upon the bordering cliffs and crags were
painfully manifest. The Sound is now actually making, and
without question it has been in this condition of self-formation
for a long time past. The ledges, islands, and isolated rocks
all speak of destruction, and of the former occupancy of the
basin by dry land. At the present time the general depth of
the Sound may be taken at about 45 to 60 feet, although our
line frequently sounded 12 fathoms; the greatest measurement
was made in what is known as the Devil's Hole opposite
Peniston Point, where the line ran out 14 fathoms. I was
informed, however, by the American Consul, the late Mr. Allen,

that soundings had been obtained in the same place of 16 fathoms. The dredge-net usually brought up from these greater depths only a grayish-white mud or ooze, largely made up of coral and coralline fragments and the debris of the crumbling cliffs. among which the perfectly formed tests of a limited number of Foraininfera—Globigerina, Orbiculina— could be made out. A deposit is manifestly accumulating on the floor of the Sound, and at a rate evidently much more rapid than that which marks disappearance through solution.

Our journeyings through the country were largely made by cart, a ramshackle two-wheeled arrangement which we canopied so as to protect us from the force of the sun's rays. That a party of nine, men and women closely huddled together, with an arrangement for traveling such as we had, should have attracted some little attention, or even drawn out the smiles of the kindly-disposed natives, goes without saying. We found it impossible at the Flatts to obtain a two-horse conveyance of any description, consequently we were compelled to put up with a simple cart, or with that in combination with another vehicle. Fortunately, the excellent condition of the country roads rendered traveling even in our rude contrivance fairly comfortable, while the load was not over burdensome to the single animal. The statement that has gained currency that two-horse conveyances are practically unknown in the Bermudas has nothing to support it.

We did not suffer so much from the glare of the roads as we had anticipated. The anticipatory warnings concerning green-umbrellas and black-goggles had succeeded in thrusting these articles of defense upon us, but they were barely more needed here than in any other limestone region. Nor did we find the heat of the sun to be of that oppressive quality which report made it. The highest marking of the thermometer during our sojourn (July) was 84° F., considerably lower than what we should have expected, during the same season of the year, for the region about Philadelphia or New York. We found but little difference between the temperature of night and day—

some four or five degrees—and usually the early hours of morning were the most oppressive. At that time the atmosphere is more settled, and in a measure lacking in those refreshing breezes which make the climate, despite the heaviness which a moisture-laden atmosphere brings with it, pleasantly bearable. Except in localities where you are temporarily debarred from the benefits of the breezes, the heat is in no way oppressive, and on the open waters we found that the difficult work of dredge-hauling could be done without special fatigue, and without drawing perspiration. Indeed, this work was not nearly as trying as I found it two years before in the waters of Nantucket Island. The balmy air of evening and the later hours is delicious, and permits of a night being passed in the open air without discomfort. Only from sudden showers is any annoyance to be anticipated. These, however, are sometimes very sudden, and seemingly the rain falls at times when it would be least expected. It was a common saying with us that a clouded sky could be relied upon, whereas the opposite was threatening. The passing off of a shower is, however, just as rapid as its beginning, and often the whole rain was over in a few seconds. Only once did we experience a lasting furious rain, but then the water descended in torrents. But within an hour after the close even of this rain the roads were practically dry.

II

THE NORTH ROCK.

In the open ocean, north of Flatts Village, lies a cluster of rocks upon which the foot of man has thus far but rarely trod. Gray and weather-beaten, they are yet firm as of old, and bear well the marks that a struggle with the sea has impressed upon them. During some six hours of the day these isolated rock pinnacles, of which the largest barely exceeds the double-height of man, are united to one another by a species of organic or living basement, while during the remaining hours they are immersed in the blue coralline sea by which they are everywhere surrounded. Nine miles distant lies Bermuda— or more properly, the hundred or more islands and islets which together constitute the Bermudas—a soft line of purple stretched against the southern sky. To the southwest the eye detects the white shaft of Gibb's Hill Light, a giant pillar capping one of Bermuda's greatest elevations—245 feet—while to the southeast the pharos of St. David's, the all-guardian of the archipelago, plays hide-and-seek with the foot-hills that nestle at its base. Beyond is all sea—the green-blue ocean in whose bosom are locked the treasures of an unseen world.

This fragment of a universe is practically all that is to be seen of the great outer reef, which lies buried, even at low water, at some little depth beneath the surface. The distance from the main-land renders access to it difficult, and it is only under exceptionally favorable conditions of water that it can be approached with advantage. Even after the surrounding shallows have been crossed it is not yet easy to effect a landing,

and we found that our pilot was little inclined to risk his boat
in the ragged growth of millepore and serpula which everywhere forms the superficial portion of the growing reef. This
inaccessibility, doubtless, accounts for the fact that so few
among the visitors to these distant shores—or for that matter,
even residents—have visited this remarkable spot, which is, beyond comparison, the most interesting feature which the Bermudas have to offer. To those who have seen the reef merely
by sailing over it, it can be well said that they have but half
seen it—they have missed its greatest charms.

The traverse of the inner waters between Flatts Village
and the reef is of itself replete with interest. Here and there
glimpses of the bottom reveal wonders of a natural fairy-land
which bid welcome to a realm of indescribable beauty. Corals
of bright orange and yellow, sponges of black and cardinal,
nodding sea-fans of purple and silver, and fishes of all that
brilliancy of coloring which distinguishes the ichthyic element
of the coralline seas, these and much more are the pictures
that appeal invitingly to a habitation in the oceanic waste.
The ruffled surface of the water bars out that clear vision
to which we are accustomed in our meadow-wanderings, but
the magic of a few drops of oil, or the stilling of the water-
glass, brings out the relief in the most wondrous detail. The
bottom bristles with a forest of rising stems and branches, the
work principally of that most indefatigable hydroid-coral, the
millepore, and through it are scattered the roses of the deep.
Countless black sea-urchins (Diadema) lie quietly nestled in the
maze, while here and there, where the animal shrubbery has
permitted the white sand to come to view, we catch passing
glimpses of the lonely black sea-cucumber (*Stichopus diaboli*),
quiet and motionless, as in the stiller waters of the Sound.
One of the most beautiful objects of these waters is the pink
tunicate Diazona whose long stems we hooked up in association
with one or more forms of Gorgonia.

The water shallows, and we approach the boundaries of the
outer reef; the huge brain-corals (Diploria) rise to within four

or five feet of our keel, and show their cerebral convolutions
with the distinctness of cabinet specimens. But those of us
who are accustomed to the white corals of museum-shelves and
mantels see nothing of that description here. The internal
framework or skeleton is completely covered by the living
animal substance, a thin film of organic jelly of the most brilliant orange in this instance, from the surface of which protrude
the ever-busy polyps. To conceive that these huge blocks
everywhere scattered about, three, four and five feet in diameter,
should be the silent work of these tiny organisms! But how
weak is the conception compared with that which recognizes in
the architecture of all the Bermudas principally the labors
of the coral animal!

Our launch is now fairly within the reef; we anchor, and
take to the whale-boat, determined to storm the little spot that
nature had bequeathed to the ocean wave. We toss gently
over the inflowing billows, and at first it would seem as though
our enterprise were to terminate in failure. But a moment
more, and success is achieved. The lee-side of one of the
massive outgrowths of millepore and serpula permits us to enter
safely into our little port, and, taking the necessary precautions
to land where the solidity of the marginal growth promised
security from a too sudden plunge into the sea, we disembark,
critics might say, in not very orthodox fashion.

To those who have never seen a growing coral-reef it is
impossible to describe the magnificence of the scene. With rapt-
urous delight and wonder you gaze through the crystal waters,
and follow the infinitude of form and color that everywhere
surrounds you. The eye rests but for a moment on one object,
it is immediately called to another. Corals, sponges, squirts,
lime-secreting Algæ (nullipores) are welded together into one
vast wilderness of coloring, a carpet mosaic of the most bizarre
pattern and brilliancy. All animal life is out in holiday
attire; the crabs, the shells, the worms are painted with the
same brush and palette which were used in frescoing the corals
and sponges. Red, green, yellow, and purple blotches appear

everywhere, and so finely are they interwoven that the outlines of the elementary parts lose themselves in the body-mass of the whole. Thus has nature provided her weaker offspring with a protective coloring, and allowed them to live almost unobserved amid the haunts of their enemies. We found the *Diadema setosa*, the peer of all sea-urchins, very abundant on the reef, where in magnificent contrast to the wealth of color by which it was surrounded, its ebony-black masses stood out in prominent relief from the coral shelters which it inhabits. All the individuals occupied recesses in the coral growth, which they had by some means, probably, managed to keep open. It is a noteworthy fact that while most of the animal forms inhabiting this portion of the growing reef were brilliantly colored, harmonizing with, and shielding one another by, the party-tints that had been culled from the rainbow, these urchins were alone conspicuous by the absence of any such protective cloak; but just in their case no protective guise in the form of coloring is needed, since these animals are abundantly able to shield themselves by means of their greatly attenuated spines. We found three other species of sea-urchin, *Echinometra subangularis*, *Hipponoe esculenta*, and *Cidaris tribuloides*, fairly abundant in the coral shelters, the last-named species, I believe, being now for the first time recorded from the islands.

We secured many specimens of the large Diadema for our collections, but found that caution in handling was necessary, lest the needle-spines would be projected into the flesh, and there broken off in fragments. In what precise manner the animal so deftly manages to insert its spines into the finger tips, and leave them there as reminiscences of its existence I could not determine; but the fact spoke of an occult operation in painful language. The urchins are, however, not the only animals that produce lasting impressions upon the visitor to the reefs. The corals and millepores are all endowed with stinging powers, and the ulcerations and swellings which their nettle-cells produce are frequently retained in quiet discomfort for several days. The jelly-fishes and Medusae are equally disagreeable in their companionship, and on two occasions we found that long after stranding, and for hours after life had been apparently extinguished, the Portuguese-man-of-war was still able to discharge with effect its tiny darts, and produce an impressive irritation.

Of the larger jelly-fishes frequenting the neighborhood of the reef we found the pink Aurelia and the rhizostome especially numerous, and it was interesting to watch with what equability these delicate creatures weathered the rolling billows, how nicely they adjusted their presence so as to evade the breaking water, and kept their pulsating bells in the comparatively quiet zone of only a few inches depth beneath the surface.

The more tranquil pools of the reef are in places stocked with fish, which partake of the same brilliant mould which distinguishes the lower animals. The members of the tribe of labroids, such as the golden "Spanish lady," the "blue-fish," and "hind" were especially conspicuous, darting rapidly from shelter to shelter as our shadows were cast upon the water. Wading through one of these pools I disturbed the peace of some thirty or forty fishes of the most intense blue and purple, but the rapidity of their movements rendered a determination of the species impossible. We observed, however, none of the lovely angel-fishes, with their long streamers of blue and yellow, nor any of the parrot-fishes proper, which apparently find a more congenial home in the inner waters of the archipelago.

The surface of the reef that is here exposed above low-water is barely more than a few yards in width, and is everywhere honey-combed into pits of greater or less depth. Many of these pits were dry, or nearly so, while others are permanently filled with water; but whether you examine the one or the other, you find the same traces of a busy animal life. Tiny crabs are especially abundant, and they may be seen scurrying about in all directions; as elsewhere the hermits have well provided themselves, and the moving domiciles of Purpura, Nassa, and Columbella, with their colored patches of algal growth, are conspicuous objects on the floor of the reef. Seemingly none of the larger or more conspicuous shells of the archipelago are found here, nor indeed, can it be said that shells of any description are numerous.

The predominant form of coral, at least in the upper layer, is the Porites, whose masses of bright orange appear here and there through the more sombre tints of the serpula by which they are almost everywhere covered. It grows to within a few inches of the water-line, but nowhere did we see it protrude above the surface, even at lowest water. This is true of all the stone-corals with which we came in contact, and also of the millepore. But we found large encrusting patches of that singular actinioid form, *Corticifera flava*, completely exposed, and beyond the reach of spray. The length of exposure is, however, short, and probably sufficient water is retained during this time within the cavernous mass to minister properly to the wants of the organism.

The serpula grows in dense bunches, forming a true basement, which is admirably adapted toward withstanding the attacks of the sea. Indeed, everywhere along the border where the surf beats hardest, the serpula growth was most largely developed, and to such an extent as to form a raised rim or barrier to the more protected inner side. Breaking in on all sides the surf has created a number of more or less irregularly oval islets with depressed centers—or, more properly, with elevated borders—diminutive atolls, as it were; and, indeed, this structure has led naturalists to assume that the form of the true coral atolls, with their central lagoon and bounding outer ring, may have been produced in much the same way, and without the assistance of any such subsidence as was considered necessary for their formation by the late Mr. Darwin. I feel satisfied, however, that the two structures, while seemingly alike, have practically little or nothing in common; in the one case the central depression is merely a negative one, being such by reason of a somewhat more rapid growth developed only from the water-line, or within the surf; while in the other, the hollow extends frequently to depths far beyond the zone of coral growth, and where, consequently, the building force is entirely removed from the influence of special conditions existing in the superficial layers of the water. We may not yet have fathomed the true method of the formation of coral islands, but such evidences as I was able to obtain at the Bermudas failed to convince me of the erroneousness of the time-honored theory of subsidence which was first formulated by the illustrious Darwin, and which has so long ministered to the wants of the geologist, and still more failed to satisfy me with the demands of the younger school of geologists, who, under the lead of the venerable Duke of Argyll, have pinned to their standard the now almost classical motto: "Conspiracy of Silence."

I could not readily determine to what extent the ocean side of the reef was more profuse in its coral growth than the inner side. Seemingly there could not be much difference, for the profusion of the inner life was such as to make it difficult to conceive how it could have been measurably exceeded. Probably in this regard the Bermudas form an exception to the supposed general rule which defines a comparatively barren area immediately back of the surf. Surely, we found nothing of the kind here.

I was also much impressed by the fact that there were here no loose boulders of rock, such as it might have been assumed would be thrown up by the disintegrating force of the breakers. Everything was firm and compact, except along the margins, where the growth of millepore formed veritable, but readily destructible, *chevaux-de-frise*. Walking on this part is dangerous, since it is not always easy to determine how strong the growth is, nor how soon one may find his way into one of the numerous water-passages which honeycomb the mass. We had experience of this danger in wading within the millepore shallows of Devonshire Flatts. The absence of coarse debris is, doubtless, in a measure, due to the small extent of land exposed, and to the depth of water which covers the greater part of the reef. A rise in the reef would probaby bring about other conditions—as was manifestly the case formerly—but even then the solidity and compactness of the growth would render the process of undermining and disruption a slow one. This is also true of the reefs on the south side of the island, the crests of which are serpuloid, while the lee-slopes are fairly covered with large Mæandrinas. We but rarely came across a loose block of stone on the beach, and where such was found it could be generally, if not always, identified as the disrupted part of the cliffs upon which the fury of the surf was expended. The examination of the lime-sand of the inner waters only exceptionally showed recognizable coral fragments, although it was very largely composed of the debris of the more friable millepore. Indeed, it might be said that the sand is properly a shell and millepore sand, rather than one of coral formation, and this is also true in a measure of the cliff-sand of the main body of the islands.

We passed the better part of three hours on the reef, but by the end of this time the water was gaining upon the spot rapidly. In a few short hours the reef would again be entirely covered, save the three gray pinnacles which constitute the lone North Rock.

III

THE PHYSICAL HISTORY AND GEOLOGY OF THE BERMUDA ISLANDS.

The reefs, islands, and lagoons which together constitute the Somers Archipelago cover an elliptical area, bearing somewhat east of northeast, of about 220 square miles, of which by far the greater part is occupied by water. The land portion is confined almost wholly to the south and southeast, where it makes a broken irregular crescent, some fifteen miles in length, and from one to three miles in width. Five principal islands, of which the largest, with nearly 10,000 acres, contains approximately three-quarters of the entire land surface exposed within the archipelago, are the components of this crescent, about which are scattered some two hundred or more islets and isolated rock-pinnacles. The great body of water or lagoon, as it is sometimes called, which lies north of this chain of islands, and is in direct communication with the open sea, is in a measure delimited by the ellipse of the outer reef, which is wholly submerged even at low water, except at two or three points, the most conspicuous of which is at the North Rock.

The depth of water in this outer lagoon is, in general, about 40–50 feet, although, exceptionally, our sounding-line dropped to 10 or 11 fathoms. For some little distance before the outer reef is reached the water shallows to 20–30 feet, and at various spots within the open, serpula and mille-pore growths rise to within a foot or so of the surface, or even completely up to it, forming irregular oval patches, which can be distinguished even at a distance by the discoloration of the waters.

An open continuation of this outer water is the Great
Sound, which is to an extent land-locked by the "hook"
of Main Island, and its disrupted continuation, Somerset
and Ireland Islands. At the eastern end of the archipelago
an incursion of the southern waters has formed, or helped to
form, the lagoon known as Castle Harbor, an extensive
body of water, with a depth of from 30 to 40 feet, whose
oceanic boundaries are well seen in Cooper's, Castle and
Nonsuch Islands, and their dissociated fragments. Castle
Harbor stands also in direct communication with the northern
waters by means of one or more channels, known as "The
Reaches," which are in part largely silted and coral-grown, and
consequently difficult of passage.

Harrington Sound, which bites into the Main Island alone,
is the most nearly land-locked of the inner waters, and at the
same time the deepest water in the archipelago. The average
depth is probably not less than nine or ten fathoms, and our
line frequently dropped to 12 fathoms. We sounded 14
fathoms in the southern bay opposite the Devil's Hole, and I
was informed that 16 fathoms had been obtained in the same
locality. Although two miles in length, and approximately
a mile and a half in greatest width, this extensive body of
water communicates with the outer sea by a channel not more
than 50 feet in width, the Flatts Inlet. The breaking action
of the waters, the undermined ledges, and the vertical cliffs all
clearly indicate that the Sound is still expanding, and it is
merely a question of time, if the present conditions continue,
when it will be more in the nature of an open bay than of a
land-locked lagoon.

The reefs on the south side of the archipelago approach in
places to within a hundred yards or as many feet—or even less—
of the chain of islands, from which they are separated by a
belt of water of no inconsiderable depth, and always existent.
They are, therefore, more nearly in the nature of "barrier"
than of "fringing" reefs, if, indeed, they can be said to strictly
belong to either one of these two divisions. Opposite the open

way communicating with Castle Harbor we sounded nine
fathoms in the water back of the reef, and I believe that
this depth, or even a considerably greater one, must be found
in many places. On the outer side the depth of water increases
more rapidly, but not in a degree as to indicate abruptness of descent. The organic growth, which is serpuloid superficially, comes to the surface in discontinuous patches, over
whose line a white surf may be seen breaking during the
greater part of the day. These are the " boilers," or secondary
atolls, as they have been sometimes called.

The outer soundings made by the officers of the " Challenger"
indicate a gradual descent of the bottom for a distance of
about a mile, when a much more abrupt slope begins. It is
claimed that within the first belt the average depth does not
exceed 12 fathoms, but we sounded 13 fathoms, after making
full allowance for slip, at a distance of not more than 300 feet
from the breaking surf. Our facilities, however, did not permit us to make extended observations in this direction, nor
was the condition of the water, when we crossed over the reef,
favorable for such observations. At a point seven miles from
the northern reef the "Challenger" obtained the extraordinary
depth of 12,000 feet, which would give an average descent of
slope of one in three, exceeding probably that of any equal
slope on the land surface. Drained of its surrounding waters,
the Bermudas would, from this side, appear like a stupendous
tower, in comparison with which the principal mountain
peaks of the land would, as far as abruptness of slope is concerned, sink into insignificance. Yet it would seem that, in a
general way, at a distance of ten miles in nearly all directions
the depth is only from 9,000 to 13,000 feet, or not more than
that which is found at an equal distance off the sub-continental
plateau. Twenty miles to the southwest-by-west of the Bermudas two submerged banks come to within 20–50 fathoms of
the surface, and seemingly represent the continuation of a
range of heights of which the Bermudas constitute a section.
But not impossibly they are merely the summits of isolated

volcanoes, such as the Bermudas may themselves be; the distance between the two groups is amply sufficient to allow of both of them to slope gradually and independently to their bases without necessitating the assumption of a connecting backbone or ridge. The great depth of water, moreover, which lies at no great distance to the west, and likewise in the east, would seem to offer no support to the notion of such a submerged ridge, which would necessarily have to be of limited extent. Still, the shortness of the line cannot be looked upon as strictly negative evidence, since abbreviated chains with lofty summits are not absolutely unknown, even if they are of exceptional occurrence.

The main islands of the archipelago present approximately identical features. Gently undulating hills, rising sometimes with the symmetry of sugar-cones, alternate with broadly open lowland, and pleasantly diversify the landscape. Along much of the northern shores these elevations gracefully descend to the water-line, where they form long reaches of sand-beach, or terminate in abrupt escarpments, largely undercut, and usually of inconsiderable height. Viewed from an eminence, this succession of undulating hills and dales, or perhaps more properly stated, "ups and downs," with their inclosed lagoons, projecting promontories, and scattered islands and islets, forms a most captivating picture, whose beauty is further enhanced by the magnificent contrasts of color that are presented. To the geologist the picture immediately suggests a region of submergence, or such as would be formed were the more interior districts of Main Island suddenly depressed beneath the water.

At certain spots, well shown on the northern and southern shores of Harrington Sound, and on the Walsingham tract of Castle Harbor, the water has cut vertical faces from the hill-slopes, and constructed cliffs of majestic and picturesque appearance. The Abbott's Cliffs of Harrington Sound have an altitude of probably not less than 50 or 60 feet. Along the south shore a long line of almost continuous and imposing cliffs faces the ocean. These receive the full force of the battling waters, and are cut and moulded into ragged masses wholly unlike anything that is to be found on the opposite shore. This picture of wild magnificence—the beetling cliffs and dashing spray—is a surprise to the stranger who has conceived of the Bermudas only from the north, and wanders over to this side expecting to see the picture with which he is familiar repeated. Long before the shore is reached the character of the work that is here being accomplished can be judged of from the continuous booming that falls upon the ear. Deep bays, alternating with bold and ragged promontories, bite through the cliffs in some places, while at others they are still in course of formation. Just west of Hungary Bay and at the Chequer Board, perhaps the grandest views of destruction are presented, but almost everywhere the picture unfolds itself in much the same detail. We could, however, form no true conception of the destructive power of the sea from the condition of the water at the time of our visit to the islands. In the season of storms, and more particularly during a hurricane, the fury of the waters must be almost irresistible, if we give full credence to the reports of experiences of the inhabitants; and the landmarks that the sea has impressed upon the country leave no room for doubt as to the authenticity of these reports. The natural arches at Tucker's Town, which are now not even reached by the sea, bear ample testimony to an extent of erosion which is not permitted to the sea in its ordinary condition; and the same is true of the massive cliffs, some 80–90 feet, or more, in height, which constitute the "amphitheatre" a short distance beyond the arches. Several considerations preclude the probability of these structures having been formed at a time when the relations of the land and water were different from what they are now, or that changes of level have taken place since their formation. The evidences of recent encroaches of the sea at these points are clearly visible, while there seems to be nothing to indicate a late rise of the land-surface. Still, I must admit that the observable facts at our command were not sufficient to warrant us in assuming positively that there was no such elevation ; but the reverse could just as well have been, and more likely was, the case.

The loftiest eminence in all the Bermudas is Sear's Hill, about a half mile southeast of Flatts Village, which attains the modest height of 260 feet. We verified barometrically the earlier determination of this height. After Sear's Hill, the highest point is reached in Gibb's Hill, 245 feet. There are no ponds, springs, nor flowing-bodies of freshwater throughout the archipelago, although at one or two points the water of interior collecting pools is only feebly brackish. In a cattle cistern or spring, near Peniston Pond, there was little or no salinity appreciable, although the water did not appeal invitingly to the human gustatory sense. The porosity of the rock almost immediately absorbs all falling water, and likewise conducts the sea-water into the innermost parts of the islands, where it doubtless forms a clearly defined basal zone. Much of it must be drawn by capillarity above sea-level. All attempts to obtain freshwater by means of artesian borings have resulted in failure, by reason of the complete permeation of the oceanic waters. The large interior ponds or lakes, all of which occupy low levels, are necessarily brackish, and they support a fauna distinctive of brackish or salt waters. A fairly large peat-bog occupies the center of Main Island, and apparently marks the site of an ancient, now wholly desiccated, lagoon. The peat is said to extend down to a depth of 40 or 50 feet* below the sea-level, or to about the level of the floor of the great outer water which is inclosed by the northern reef.

The rock of the islands is a granular limestone, which is in most places still sufficiently soft to permit of being cut by a hand-saw. On exposure to rain it hardens through cementation, or deposition of interstitial lime, and may then be converted into a tough and very resisting material, which is advantageously used in the construction of houses. Piles of hand-sawn blocks awaiting induration are a not uncom-

---

*Rice: Geology of Bermuda, Bull. U. S. National Museum, No. 25, p, 7. On the authority of General Lefroy.

mon sight along the roadsides.   The process of the binding together of the loose particles of debris which are to constitute a rock is sometimes a very rapid one, especially along the water's edge, and may be followed in its different stages.

The basal rock of the cliffs, especially on the south shore, is in places excessively indurated, and about as resisting as a non-siliceous limestone can well be; when struck with a hammer it at times rings with all the intensity of the volcanic ringing rocks, and chips off as sharp-edged flakes.   The granular structure which is so prominent in the softer rock may be retained, but it sometimes largely or wholly disappears, and the mass appears to be homogeneously compact.   The matter of hardness or compactness is, however, not one necessarily depending upon age, since we often find the tougher rock occupying the high level, and overlying the softer rock below. At other places the two kinds of rock alternate with one another.   From the constancy of its occurrence at, or near, the base of the island, the hard subcrystalline limestone is locally known as the "base rock;" it serves largely, but by no means invariably, to distinguish the old beach formation, and thus to locate the former sea-border.   The same rock forms the lower moiety of the three pinnacles of the North Rock.

The constituent particles of the softer limestone are of about the size of a pin's head, or smaller, among which the debris of shells and millepores are distinctly recognizable.   Coral fragments are apparently much less abundant, and, indeed, it was only with difficulty that I determined these at all, except where, at long intervals, fragments of large size were caught up in the mass.   Possibly, the finer undefined particles may have been those of corals, whose cellular structure would have readily lent itself to a powdering such as would completely efface determining characters.   Still, the fact remains that much, or most, of this rock is made of millepore and shell fragments, and whatever part corals may have taken in its formation, it cannot be considered to be a simple coral rock. The examination of the long stretch of beach which faces the north side of St. George's Causeway also failed to show much evidence of coral growth, although shells and millepore fragments were packed in endless quantities; the tests, perfect and imperfect, of the foraminifer genus Orbiculina were also very abundant. I do not wish to be understood as saying that the islands are not really of coral formation; that a coral fundament exists, needs no further demonstration than is furnished by the rich growth of Diploria and Mæandrina within the reef-waters, and by the coral fragments and masses that are inclosed by the beach formation. I wish merely to emphasize the important part which organisms other than corals have taken toward the shaping and the making of the rocks, especially of the superficial parts which have lent themselves to wind-action.

The true relations of the Bermudian rock were first clearly established by Nelson.* With remarkable sagacity this observer read the history of the discordant layers, here horizontal, there steeply inclined, now arched in one direction, then in another, which appear in all the sections, and he was not slow to point out that they were the result of wind-drift—mere shifting (calcareous) sands which had been thrown about promiscuously by the winds, and had solidified in layers in the positions where they had been finally dropped. This interpretation stands to-day as firmly established as it stood when it was first enunciated upwards of a half century ago. The thin knife-sheets which are so characteristic of this drift-rock build up massive beds, which are thrown together in most irregular confusion—dove-tailed, apparently faulted, lenticulated, and otherwise. No more interesting exposures can be had than the faces of the road-walls, both in the city and in the country, where synclines, anticlines, slopes, and horizontals appear sometimes in the space of a few yards. At other places no bedding, beyond the thin lamination, is apparent, and the whole mass rests concordantly either in straight or undulating lines.

*Trans. Geol. Soc. of London, 2d Ser., vol. v, Part I, pp. 103–123.

The first process toward the forming of this rock must necessarily be the pounding up of the material out of which it is constructed. Wherever the polyps build close to the surface their habitations are attacked by the surf which they themselves create. The long white line of foam which meets the eye of the observer gazing southward from any eminence, and parts the blue waters of the outer world from the more nearly green within, is but the line of battle between the organic and the inorganic forces. It is here that life asserts her supremacy over the sea, and it is here that the sea maintains her right of domain as an inheritance of prior birth. Blocks of coral and coralline are detached and broken, their parts are rocked to and fro in the withering crest, and ultimately, when the fragments have been sufficiently punished by the sea, they are handed over for further chastisement to the action of the wind. In this way the particles are ground finer and finer, a true sand is formed, and dunes begin to rear their heads above the ocean level. Traveling in the line of the wind the dunes pass onward, climb over one another's backs, and comb the gently flowing crests; from pigmy hillocks they rise into well-fashioned knolls, and ultimately stand as the eminences which to-day are the Bermudas. No one who, on the south shore, has watched the great tongues of moving sand,—the sand glaciers of Tucker's Town and Elbow Bay, for example—stealthily encroaching upon the hill-tops of the interior, and burying everything, in the manner of the locusts of South Africa, beneath their mantle of destruction, can have failed to be impressed by the character and the magnitude of the work that is being accomplished. It is truly but the music of the sea and wind, but there is enough of it to turn water into land. It seems, indeed, as though Virgil had divined some such region as this when he depicted the home of Æolus in the following beautiful lines :—

> Here Æolus, in cavern vast,
> With bolt and barrier fetters fast
> Rebellious storm and howling blast.
> They with the rock's reverberant roar

Chafe blustering round their prison door:
He, throned on high, the sceptre sways,
Controls their moods, their wrath allays.
Break but that sceptre, sea and land,
And Heaven's etherial deep,
Before them they would whirl like sand,   '
    And through the void air sweep.

(Conington's Æneid.)

The æolian or wind-drift character of the Bermuda Islands is everywhere apparent; along the roads, on the hillsides, and in the caves we find the same rock made up of organic particles. The layers or seams, inclining now one way, now in another, point to the different positions into which the sand had been fortuitously cast by the winds, patted down, and built up into a series of superimposed layers. Shells, both marine and terrestrial, have been caught up in the drifts, for we find them now embedded in the rock, and scattered over the most remote corners of the island group. I picked up a fairly large fragment of coral at an elevation of probably not less than 150 feet; and, doubtless, other equally large fragments occur at still greater heights. In regions where freshwater streams abound, the materials of terrestrial destruction are washed into these streams, and by them carried' into the sea; geologists have long since recognized the force of the saying: "the land-surface is on one grand march to the sea." But here, where freshwater streams are entirely wanting, and the falling drops are immediately absorbed into the porous soil, the conditions are at least in one sense reversed—the march is in a direction away from the sea. Whither it may ultimately tend cannot be foretold.

It is difficult to conceive that these lovely hills, buried beneath their sombre covering of juniper and sage (*Lantana*), should have been thus shaped by the wind; but the facts are plain in their statement, and leave no loop-hole for the doubting mind. The height of the sand-hills, or dunes, for such they really are, is unusually great for a coral island, and serves to

distinguish the Bermudas from other islands having an apparently related structure. I fully concur in the suggestion thrown out by Prof. Rice that these accumulations could only have been formed at a time when large areas of reef, and not a simple atoll-ring, were exposed above water-level. At the present day nearly all the sand is formed through the destruction of the existing land-mass, and not as a product of disintegration derived from the growing reef.

Prof. Rice, in his interesting observations on the geology of Bermuda (*loc. cit.*, pp. 10–13), correctly distinguishes a "beach" rock as underlying in many places the drift rock of the shores. He instances as examples of such rock the fossiliferous stratum which appears in the chain of islands stretching across Hamilton Harbor, the conglomerate of Stock's Point, near St. George's, which rises some twelve feet above the water, the lower bed of Devonshire Bay, and much of the basal, nearly horizontal, strata which appear on the south shore. As characteristic of this beach rock, it is said that the beds nearly uniformly dip seaward, but at a very moderate angle, and that they contain largely of the remains of marine animals (corals and shells). The rock is in most cases very tough and hard, and is largely the correspondent of the base-rock that has already been described.

We also found this beach-rock well developed at many points along the south shore, where it rises some 5–8 feet, or exceptionally more, above the sea-level. The series of nearly horizontal ledges, sharply defined by their position from the highly inclined layers of drift-rock by which they are surmounted, or into which they graduate landward, which appear basally at the Chequer Board and at Harris's Bay well illustrate the characteristics of the rock. I failed however, to detect the uniform seaward slope of this rock which Prof. Rice indicates, nor could I satisfy myself that the presence of marine organic remains in a rock were conclusive for considering the rock to be of beach formation, unless, indeed, such remains were abundant, or else showed by their positions that they could only have been placed there through the normal manner of oceanic deposition. It is true that in by far the greater number of cases the rock that can be identified as of beach formation contains, when remains are present at all, only the relicts of marine organisms, and that the drift-rock above or back of it contains only the parts of land mollusks. But Prof. Rice justly remarks that the remains of land organisms can be readily washed or drifted into the sea, and there combined with the organisms that are subsequently to enter into the formation of a beach-rock. A mixed faunal element would thus be introduced. But much the same kind of intermixture may take place in the land-deposits through the washing or sweeping on high of marine organisms, or their fragments, especially during periods of high storm. Prof. Rice recognizes the possibility of such intermixture, but he attributes it all to the action of the wind. It is claimed that only small or light fragments can be swept up by it, and that necessarily only these can be found, under ordinary conditions, drifted into the rock. A fragment of the shell of Spondylus weighing 1.8 grammes, a valve of Chama, incrusted with tubes of serpula, weighing 2.7 grammes, and a fragment of Mycedium, weighing 8.3 grammes, were found by that investigator in the sand-drifts of Tucker's Town, and these weights or masses are given as values of the carrying power of the wind. This, it appears to me, is doing scant justice to the assistance which the wind receives from the sea. Under ordinary conditions the action of the sea may be confined almost wholly to the line of beach, but it certainly is otherwise during storms. At such times there can be no question that much in the way of organic remains is thrown far within the domain of the drift-rock. The hurling of pebbles and stones along exposed coast-lines is sufficient evidence of the capabilities in this direction. We were given graphic accounts of the violence of the waters under exceptional conditions of storm, and were shown, in the house of the Misses Peniston, at Peniston's, a position reached by volumes of spray which we should have believed impossible, were it not for the absolute reliability of the residents of the house who volunteered the information.

At several points more particularly along the north shore I found marine shells (Lucina, Tellina, etc.) imbedded in unquestionable drift-rock, and, indeed, it could hardly have been expected that such association should not occur. On the whole, however, these remains were not as abundant as one might have expected to find them. The same is also true in a measure of the occurrence of land-snails. One of the commonest shells of the lower drift-rock is the large *Turbo* (*Livona*) *pica*, a shell which appears to be very abundant about the coast, but which generally, and perhaps invariably, is cast up without the animal. I was unable to find anyone among the local collectors who had seen the animal itself, nor did any member of our party succeed in obtaining an occupied shell. Nelson and Rice both attribute the occurrence of this shell in the drift-rock to transportation by hermit-crabs. I can hardly believe that this is the full explanation. I failed to find any of the shells of the beach inhabited by hermits, and was in this respect less fortunate than Nelson, nor do I know of any hermit of the islands which would be likely to carry about with it the massive full-grown shell. However, my testimony on this point is purely negative.

I admit with Prof. Rice that it is frequently difficult to distinguish between what is assumed to be beach-rock and the regular drift-rock of the islands, especially when the latter occupies a basal and nearly horizontal position. In many places along the south shore where the beach-rock is exposed in heavy beds it occupies but a limited horizontal space, being succeeded by highly inclined drift-rock, which descends to the water-level. This succession is unexpected, and might lead one to infer that there have been local differential movements on the part of the land. But of course this need not have been, and doubtless was not, the case, since an irregular or indented shore-line undergoing elevation would form features similar substantially to those which are here presented. That is to say, the raised beach-line would be an interrupted one—continuous possibly along an inner contour, but broken on the outer face, where a low-level beach would mark that portion of the shore which had last risen. In this way, it is not improbable that much of the interior drifts of the Bermudas will be found to be underlaid by elevated beach-rock, and that a continuity of extent actually exists. It appears to me that geologists have not taken sufficient account of the irregularities in an ascending coast line as factors determining the positions or relative altitudes to which points of elevation must necessarily attain. They are too ready to interpret the obliquity or inclined position of marine terraces on the assumption of terrestrial oscillations.

The one fact above all others which immediately appeals to the geologist in the Bermudas is the rapid waste which the islands are undergoing and have undergone for some long past period. Everywhere along the coast we have evidences of this waste; the outer cliffs, the cliffs and ledges of the inner waters, the serially disposed islands and islets, all bear witness to a common annihilating process. Along the south shore the lesson of destruction is presented on the most impressive scale, and it is here that we read most clearly the record of waste which the islands have undergone. The huge cliffs are still being undermined and are still crumbling, but they are merely the remains of a land-mass that at one time projected far beyond the present coast-line into the sea. This is clearly shown by the disposition of the drift-rock of which they are composed, the layers of which in most places decline steeply in the direction of the land, turning their basset edges to the sea. Manifestly, the cliffs are merely the inner halves of dunes, the outer slopes of which have been carried away by the sea. The height of the cliffs indicates dunes of great extent, but it will probably never be told at what point in what is now sea they originated, and how much they have lost through oceanic erosion. Not improbably the land at one time projected at least as far southward as the position which is now occupied by the crest of the reef.

There can be no doubt, it appears to me, as Rein* first clearly demonstrated, that Harrington Sound is not the lagoon of a marginal or secondary atoll, but merely a hole that has been cut out of the land by the sea. I think that every one who has seen the working condition of the water in the Sound, the undercut ledges, the scattered islands and rocks, and above all, the precipitous cliffs, which appear on opposite sides of the water, and show an arrangement of lamination or stratification similar to that which is observed in the cut cliffs of the south shore, must arrive at the same conclusion. The same is manifestly also true of much, if not the greater part, of Castle Harbor, which still retains a sea-ward border in the belt of disrupted land which forms Castle Point, and Castle, Goat, Nonsuch, and Cooper's Islands. The widening or expansion of this body of water presents itself vividly to the eye of the observer stationed on an eminence, such as that of St. David, whence the field of vision takes in the patches of separating and separated land which are awaiting the hour of their destruction.

Along the borders of Castle Harbor—at least as far as we observed it on the west and south—there is a broad flat ledge, over which the depth of water is only from about six to ten feet; beyond this there is an abrupt drop into the deeper parts of the lagoon. This feature is frequently found in the true atoll-lagoons, where it forms a shore platform similar to that which is formed around the outer surfaces of sea-cliffs. In how far this ledge may represent a simple coral outgrowth from the shore, or determine a measure of subsidence, cannot well be ascertained. Large numbers of giant brain-corals (Mæandrina and Diploria), measuring three, four, and five feet in diameter, are scattered over it, and form a series of stepping stones in the water. Many of them grow on and over the edge of the platform, so that the latter overhangs in some places. These corals appear to be absent, or at least

---

*Bericht Senckenberg. Naturf. Gesellsch., 1870, p. 153.

largely wanting, in the deeper waters. We sounded at various points in 5-6 fathoms, and whether this represents a general depth or not, it is certain that the basin is far shallower than that of Harrington Sound.

In the pinnacles of the North Rock we have probably the most imposing lesson touching upon the annihilation of the land-mass. The lower portion of these rocks is, I believe, unquestionably of beach formation; I failed to detect in it the fossils (Cypraeas, etc.) which Rein asserts are to be found there, but possibly my search was not sufficiently systematic to bring them to light. This basal portion of the rock is exceedingly tough and compact, and rings loudly when struck with a hammer. The upper moiety is made up of distinctly laminated or stratified drift-rock, which dips at a steep angle. Manifestly, the materials of this aeolian formation must have had some starting ground, and could not have been developed from the small area which is exposed at low water about the base of the pinnacles. The height to which the well-indurated drift attains, some twelve feet or more, taken in conjunction with the vertical reduction which the rock must necessarily have undergone, and the destruction which has ensued elsewhere, argues almost overwhelmingly for considering these fragments to be merely the remains of a land-mass which had at one time very considerable extent, and not improbably actually united with the main islands. The work of destruction, according to this interpretation, may have wiped from existence a piece of territory possibly not inferior in area to that which is now exposed above water.

In view of the rapid destruction which the islands are undergoing it remains to inquire what are or were the special conditions which have permitted this destruction to take place, and have so completely reversed the history of the archipelago. For evidently the conditions under which the islands were first built up, and which permitted them to acquire their full development, must have been very different from those which are to-day bringing about only annihilation. In order to trace these changes it is first necessary to determine in how far the present outline or area of the Bermudas is a permanent one, or in how far it may have varied during the period of its existence. By geologists, generally, the island group is considered to represent the disrupted parts of an atoll-ring, most of which (as is seen in the northern reef) now lies submerged beneath the water. This is the view which is upheld by Dana in his "Corals and Coral Islands" (p. 218) and by the late Sir Wyville Thomson in his work on "The Atlantic." The latter states* that the character of the Bermuda atoll "is much the same as that of like reefs in the Pacific, with certain peculiarities depending upon the circumstance that it is the coral island farthest from the equator, almost on the limit of the region of reef-building corals." The atoll character of the island group is also conceded by Prof. Rice, but this authority carefully distinguishes between the present outlines and those which belonged to the original atoll; he recognizes movements of elevation and subsidence, which have practically obliterated the normal form of the atoll, and have left it in a condition where there need be no necessary correspondence existing between the present land-masses, with the submerged reef, and the primary atoll-ring. The condition is thus stated by him: "The series of movements required to account for the main features of Bermudian geology seems to be the following: 1. A subsidence, in which the original nucleus of the islands disappeared beneath the sea, the characteristic atoll form was produced, and the now elevated beach-rock was deposited. 2. An elevation, in which the great lagoon and the various minor lagoons were converted into dry land, and the vast accumulations of wind-blown sand were formed, which now constitute the most striking peculiarity of the islands. 3. A subsidence, in which the soft drift-rock around the shores suffered extensive marine erosion, and the shore platform and

*Op. cit., I, p. 302.

cliffs already described were formed.[1] With this conception
the atoll practically disappears, since, in the absence of atoll
characters, there is nothing to indicate that the structure was
ever present; at any rate, its existence is rendered purely
hypothetical.

Darwin discusses the subject with his usual perspicuity, and
finds reason to doubt that the islands are a true atoll. He
points out their close general resemblance to an atoll, but indicates the following important differences: "first, in the margin of the reef not forming a flat, solid surface, laid bare at
low water, and regularly bounding the internal space of shallow water or lagoon; secondly, in the border of gradually
shoaling water, nearly a mile and a half in width, which surrounds the entire outside of the reef; and thirdly, in the size,
height, and extraordinary form of the islands, which present
little resemblance to the long, narrow, simple islets, seldom exceeding half a mile in breadth, which surmount the annular
reefs of almost all the atolls in the Indian and Pacific oceans."
The great height of the land, as compared with other islands,
is also commented upon.[2]

In all these characters the Bermudas unquestionably differ from a typical atoll, but allowing for the conditions which
Prof. Rice suggests these differences lose much of their significance. They are not antagonistic to the notion of an overdone
atoll which is now undergoing destruction. But it is difficult,
if not impossible, to demonstrate the atoll condition itself.
If it ever existed it has been completely masked by overgrowth,
for I believe the facts such as they are show with sufficient
clearness that the present islands and reefs have little or nothing in common, beyond occupying position, with a pre-existent
ring. Matthew Jones has well argued[3] that a bodily uplift of

[1] Geol. of Bermuda.   Bull. U. S. National Museum, No. 25, pp. 16–17.

[2] Structure and Distribution of Coral Reefs, 1842, p. 204.

[3] *Nature*, Aug. 1, 1872.

some 50 or 60 feet would lay dry practically the whole archipelago, as far as the great northern reef. That such a condition of elevation at one time existed is, I believe, all but demonstrable; and if this is true the present condition can only be accounted for on one or two hypotheses: simple erosion or erosion in combination with subsidence. The vast amount of erosion that has taken place has already been referred to, and it is barely necessary to enter further into its details. It will immediately suggest itself to the inquiring mind that this erosion could not well have taken place without subsidence, otherwise it would be difficult to conceive, except under a condition of very rapid elevation, how material could have initially accumulated, so as to lend itself to destruction afterward. To assume rapid elevation, followed by a period of stability when destruction would exceed construction, requires the formulation of causes which are not less difficult to receive than those which would explain subsidence. Unquestionable evidences of subsidence are, however, by no means wanting, and coincidentally they point to an amount of movement which would account approximately for the depth of the great lagoon. Thus, in the excavations made on Ireland Island for the lodgement of the great floating dock, a deposit of peat,* and vegetable soil containing stumps of cedar in a vertical position, together with other vegetable remains, and shells of the common sub-fossil land-snail of the islands, were found at a depth beneath the water of some 45 to 50 feet. The depth of the peat-bog which occupies the central part of Main Island, has already been noticed. It seems to be a not uncommon occurrence, as we were informed by the keeper of the light at St. David's, that stumps and roots of cedars are drawn up by the anchor chains of vessels riding in the waters about St. George's.

The caves of Bermuda afford equally conclusive evidence of subsidence. Many of these now occupy a level considerably below that of the sea, and consequently receive a large in-

*Thomson: "The Atlantic," I, p. 320.

draught of water from that source. In the very beautiful Island
Cave, near Joyce's Dock, on the north shore, the interior pool
has a depth of probably upwards of twenty feet—we were informed that it was thirty feet—yet through it rose a ponderous
stalagmite, several feet in thickness. Manifestly, this stalagmite could only have been formed when the pool was not yet
existent or at a time when the floor of the cave was elevated
above sea-level. The depth of the pool, then, is a measure of
the least possible amount of subsidence, from 20 to 30 feet in
the present instance. Much the same condition is presented
by some of the other caves. These sea-grottoes are among the
most attractive features of the Bermudas, and they would,
even in regions far famed for their caves, attract attention.
The principal vaults are of fairly large size, but the connecting passages are low and contracted, rendering deep penetration difficult.

These various forms of evidence make clear that there has
not only been subsidence, but subsidence on a moderately
large scale, and in a comparatively recent (geologically speaking) period. Indeed, were we to search for evidence of subsidence alone we would not be compelled to go beyond the
simple drift-rock, which at so many points dips directly into
the sea. To what amount this subsidence may have extended
cannot in the nature of things be determined, but it appears
to have been at least sufficient to account for the depth of
water which marks the lagoon and inner sounds. Before this
subsidence took place probably the entire area now covered by
the Bermudian archipelago, and much more, were dry land,
and it was at this time, doubtless, that the great sand dunes
were elevated. The prevalence of powerful winds on the south
side would tend to elevate this side of the island, while the
opposite side, not feeling this influence in any marked degree,
would remain comparatively low and flat. In a period of subsidence the low side would naturally be the first to succumb
to the waters, and would undergo submergence long before
the elevated slopes. And this is precisely what appears to have
taken place in the Bermudas.

It becomes an interesting question to ascertain how far elevated above water-line the Bermudas were at the time when they formed a continuous island. The data that are presented for the determination of this problem are mainly of a negative character. But if a subsidence of some 50 or 60 feet can be indicated, and we still have beach-rock on the islands at an elevation of some 12–16 feet, it will be necessarily assumed that the actual uplift above sea-level was at least 60 or 70 feet, unless, indeed, the movement was not a uniform or coincident one for the entire island group. This last assumption seems, however, highly improbable. It may, again, be assumed that the elevated beach-rock was lifted since the period of subsidence, and represents the closing movement of the land. Its presence therefore need not argue for elevation beyond that which is indicated by its own highest level, some twelve or fifteen feet. But the relation of this rock to the drift-rock overlying it, and the fact that the latter in so many places drops bodily into the sea, forbid such a conception. The beach-rock is manifestly old, and long antedates the last subsidence; and for anything that can be shown to the contrary, it is at least as ancient as the lagoons and sounds, and probably much more ancient. Indeed, there is nothing that could lead one to suppose that it is not the original rock which was formed when the island first came to the surface. Although now exposed on the sea-border, it is really an interior rock, as is proved by the broad band of land which must have been removed from the seaward side of the existing cliffs.

Two questions present themselves at this stage of the inquiry. One of these has been much used of late by the opponents of the Darwinian theory of coral formations, and bears upon the formation of lagoons through aqueous solution. The second considers the amount to which a possibly cavernous condition of the island may have facilitated the work of the erosion, and permitted of the present features having been formed without the aid of subsidence.

No one, it appears to me, who has examined any of the inner waters of the archipelago can for a moment suppose that the basins holding these waters could have been formed or kept open through solution of the rock supports. Apart from the special difficulties which the Murray theory carries with it, and which will be found more extensively discussed in the general chapter treating of the formation of coral structures, the facts presented by the Bermudas are such as to immediately dispose of the theory in so far as it is made applicable to them. The material resulting from rock degradation which finds its way into the waters of the different lagoons vastly exceeds in quantity that which could possibly be removed through solution; hence we have the entire floor covered with a thick deposit of ooze, as our dredgings invariably proved, and not an exposure of bare rock as we should expect to find in a basin of solution. Organic material, largely foraminiferal, is also accumulating over the floor, and the supply of formative material from this source alone is probably fully equal to that which is removed chemically by the waters. The quantity of this basal sediment is so great that during a heavy storm, as was witnessed by Rein and others, the entire water reaching to the outer reef was rendered milky white. As regards the second question, the influence of a cavernous structure upon the erosion of the land, the facts are not readily approached. Mr. Fewkes, in a paper recently published* on the "Origin of the Present Form of the Bermudas," argues that the existing relations of the archipelago are not necessarily the result of subsidence (although he admits that the evidences of subsidence are unmistakable), but of normal erosion, assisted by the breakages which in one form or another are likely to follow the honeycombing of the rock. Caves or long passages are assumed to penetrate into all parts of the islands, and by their collapse are supposed to furnish the hollows which ultimately form the lagoon-basin. This idea is not entirely new, and was already entertained by Rein.

---

*Proc. Boston Soc. Natural History, 1888, pp. 518 *et seq.*

A number of serious objections present themselves to the acceptance of this explanation. The premises are largely, if not almost wholly, of a hypothetical character. That the islands are in a measure undermined there can be no doubt, but there is little, if anything, to show that there have been breakages of the extent which would be required by the theory. Evidences of local disruption are plentiful, such as may be found in almost every region of sinks, but as far as I can see there is nothing to indicate that basins such as Harrington Sound, Castle Harbor, or the great lagoon could have been formed, or even materially furthered, by disturbances such as the cave-theory calls into existence. The even floors of these basins argue strongly against formation through breakage, as does likewise the horizontality of the beach-rock formation. The absence of all indications of disturbance from the latter is significant. But the broader question can well be asked: How could extensive cave formations, extending 40, 60, or 70 feet beneath the water surface, be brought about without subsidence? Whence would the force of excavation be obtained? The answer might be returned: solution. But there is no more reason to assume special solution in the case of the Bermudas or in other coral islands than in any marine limestone formation.

The difficulty in the problem entirely disappears if we admit subsidence, and, as has already been seen, the positive evidences of subsidence are ample. On no other theory, it appears to me, can the waste of the cliffs on the south shore be explained. The direct evidences of subsidence, moreover, do not come from a single point in the archipelago; they are found from Ireland Island and Hamilton Sound, through the Main Island, to St. George's. And this being the case, there is every reason to assume that the area which was influenced by movements of one kind or another was not restricted to the present patches of exposed land, but extended to the submerged portions of the archipelago as well.

As the result of our researches we may express the following conclusions:

1. The present form of the Bermuda Islands bears no relation to the ring of an atoll, except in so far as the outer boundary may be more or less coincident with the boundaries of an ancient atoll.

2. The existence of an atoll in the present position of the Bermudas is not demonstrable.

3. The height of land in the archipelago was formed during a period of elevation, when seemingly the entire archipelago was a connected or continuous piece of land, extending as an oval island to what is now the bounding reef on the north and on the south. It is impossible to determine the absolute amount of elevation above the water, but it appears to have been not less than 70 or 80 feet, and it may have been considerably more.

4. The lagoons and sounds were formed during a period of subsidence which followed upon that of elevation, and is seemingly still in progress, or was so until a comparatively recent period. The great degradation of the coast-line took place at this time. It is impossible to determine the amount of such subsidence, but it was at least 60–70 feet, and not improbably very much more.

It will be seen that these results, so far as they go, are in absolute harmony with the views which Mr. Darwin entertained regarding the structure of these islands. They do not prove the correctness of the Darwinian hypothesis of the formation of coral islands, but they measurably sustain it; on the contrary, they are largely opposed to the requirements of the substitute theory which has been recently proposed. Elevation and subsidence are both shown to have marked the region in its development, and these conditions are more in consonance with the Darwinian hypothesis than with any other.

The question as to what form of coral structure the Bermudas actually are—what constitutes their fundament, and how they were built to their existing level—still remains unanswered, and possibly we may never be able to answer. But I have thought it worth while to introduce a discussion of the general coral question (the chapter following), as it has a bearing on the topic at issue.

IV.

THE CORAL-REEF PROBLEM.

Perhaps no class of phenomena has been so frequently appealed to in evidence of subsidence on a grand scale as that
presented in the formation of coral reefs. Scattered freely over
a large expanse of the oceanic surface, these structures constitute features there as distinctive and prominent as do the
mountain masses on the continents. Rising in most cases
from a deep sea, and with a limited extent, their presence, as
organic accumulations, immediately suggests peculiarities of
geological construction which are to be found in no other
form of relief. It is an ascertained fact, as has been variously
demonstrated that the conditions governing the existence and
distribution of reef-building corals (Porites, Diploria, Mæandrina, Madrepora, Tubipora, Fungia, Astræa, etc.) are drawn
within narrow limits, and that they are equally of a general
and of a special character. Broadly stated these conditions
are: A surface temperature of the water never falling below
70° or 68° F.; an absence of muddy sediment; freedom from
contact with freshwaters; the necessity, in some cases, of surf
action. Accordingly, we find that reef-structures are practically confined to the warm tropical or subtropical seas, and
that they are largely wanting in tracts where exceptional cold
currents have wedged a path into the warmer waters, or where,
as at the mouths of outflowing streams, there is a free discharge of both freshwater and sediment. To this must be added the all-important fact that the reef-building corals are confined to a superficial zone of the sea not exceeding 100 or 120

feet; beyond this depth we meet only with dead coral. In the case of the Bermudas, and in that of a few other reefs, the temperature of the water has been known to descend to 66° or even 64° F., but this condition is very exceptional.

One of the most familiar and wide-spread types of coral structure is the atoll, which acquires special development in the waters of the Pacific and Indian oceans. It consists of a more or less irregular ring of living and dead coral, enclosing within its boundaries an internal body of water (lagoon), which in many cases is kept in direct communication with the exterior by one or more connecting channels of water; the breaks in the ring corresponding 'to' these passages almost invariably occur on the leeward side of the island. The atoll, although frequently so described, is rarely of a circular form, the outline being very generally elongated and angular. In extent it varies from two to three miles, or less, in length to upwards of 40 or 50 miles; where the dimensions are very small the lagoon may be completely absent, or merely indicated by a dry depression. The breadth of the coral ring itself does not usually exceed 1000–1500 feet, or somewhat more than a quarter of a mile. In the general composition of an atoll, the following parts may be recognized : first, an outer platform of coral-rock, more or less exposed at low water, which is the correspondent of the ordinary rock platforms resulting from tidal destruction ; secondly, the beach-line proper, measuring a few feet in height, and consisting of coral sand, calcareous pebbles, and triturated shells; and thirdly, the exposed ring itself with the width as above stated, over which, more especially on the windward side, a luxuriant vegetable growth is developed. The elevation of this portion of the atoll more commonly does not exceed 10–20 feet, although exceptionally the wind-swept dunes of coral sand attain a much greater height. On some islands, not necessarily atolls, however, as Anegada, one of the West Indies, the drift banks rise to a height of 40 feet, while on the Bermudas they considerably exceed 200 feet, reaching at one point, Sears' Hill, 260 feet.

On the lagoon side of the ring the shore-platform is very commonly replaced by a gently sloping sand bottom, with or without the formation of a true beach area. Frequently there juts out from the shore a growing reef-platform (upon which the coral growth is fairly profuse), which descends with a vertical or overhanging edge to a second deeper zone of coral life. Over the bottom, which presents a more or less uniform character, coral sand and debris, calcareous pebbles, the tests of various Foraminifera, etc., are extensively distributed, forming there a sticky white or bluish paste, much like that which extends for miles beyond the outer border of the forming reef. The depth of the lagoons varies from a few feet to twenty or thirty fathoms, as we find it in many island groups of the Pacific (Paumotu, Gilbert's Group, Keeling Island, etc.). In the Maldives it exceptionally attains 50 and 60 fathoms.

As reef-building corals cannot long survive exposure to the atmosphere it is manifest that the upper limit of the growing mass will be the actual surface of low-water. In the line of the breakers, or in the shallows just beyond, the coral polyps thrive in their greatest profusion, and the almost endless variety of their forms, not less than their brilliant coloring, never fails to arouse the wonder and enthusiasm of the traveler. Prof. Dana thus graphically describes the forming island: "The reef of the coral atoll, as it lies at the surface still uncovered with vegetation, is a platform of coral rock, usually two to four hundred yards wide, and situated so low as to be swept by waves at high tide. The outer edge, directly exposed to the surf, is generally broken into points and jagged indentations, along which the waters of the resurging wave drive with great force. Though in the midst of the breakers, the edge stands a few inches, and sometimes a foot, above other parts of the platform; the incrusting *nullipores* cover it with varied tints, and afford protection from the abrading action of the waves. There are usually three to five fathoms water near the margin; and below, over the bottom, which gradually deepens outward, beds of coral are growing profusely among extensive patches of coral sand and fragments. Generally the barren areas much exceed those flourishing with zoophytes, and not infrequently the clusters are scattered like tufts of vegetation in a sandy plain. The growing corals extend up the sloping edge of the reef, nearly to low-tide level. For ten to twenty yards from the margin, the reef is usually very cavernous or pierced with holes or sinuous recesses, a hiding place for crabs and shrimps, or a retreat for the Echini, Asterias, sea-anemones, and mollusks. * * * Further in are occasional pools and basins, alive with all that lives in these strange coral seas."

This description, which is drawn from the islands of the Pacific, is largely applicable to the condition of the Bermudas. Owing to the peculiar submerged condition of the reef I was unable to determine satisfactorily to what extent a breaking surf was favorable or unfavorable to the growth of corals. At the North Rock, the only accessible point of the outer reef, the millepore growth is very profuse, and large masses of Porites may be picked out from below the capping of serpula. The same condition prevails over the Devonshire Flatts, where the surf dashes over a wilderness of atoll-like islets scattered through the lagoon. But this is not necessarily evidence in favor of advantage derived from the surf, since these seemingly more favored patches are the creators of the surf themselves, and they must have risen before assistance from this direction could have been given them. Their existence seems to prove, however, that the action of the surf is no disadvantage, a conclusion opposed to that which was reached by Bourne from his careful studies of the Diego Garcia Reef.* Along the inner slope of the reef, immediately receding from what might be called the crest, as well as on both slopes of the serpula-capped southern reef, the coral growth appears in unbounded profusion, presenting a perfect maze of millepores, gorgonias, and brain-stones.

---

*Proc. Royal Society, XLIII, 1888, pp. 453-55.

There is little or no growth of coral immediately along the south shore, doubtless due to the great quantity of sediment that is constantly being washed off from the cliffs. The rocks of Harrington Sound, on the other hand, are largely fringed with patches of Isophyllia, Siderastræa, and Millepora, while in the deep quieter waters, judging from the number of our hauls, Oculina is by no means scarce. Reference has already been made to the vast development of Diploria and Mæandrina on the projecting platform of Castle Harbor, over which the water is normally in a condition of fair stability. This condition confirms the view expressed by Bourne that the coral growth of the inner waters is much more extensive than is generally stated to be the case. A very large part, however, if, indeed, not the greater part, of the floor of the big lagoon is practically barren.

Although the zone of animal activity in a coral island ceases with the water-line, the actual growth of the island does not stop there, but is continued upward by the mechanical and vegetable forces. The destructive action of the billows carries fragments of coral-rock far above the limits of coralline existence, triturating the masses into minute surfaces, and upon this improvised soil a luxurious vegetation, whose origin lies in the seeds wafted thither by the winds, or deposited by birds, may in course of time spring up. Where the action of the breakers is greatest the coral rock assumes the greatest compactness, since the fragments and particles that are derived from the mechanical wear and tear are here firmly lodged or compacted into the spaces of the otherwise comparatively loose coral structure. On the oceanic side of the island we find shallow water—ranging to several hundred feet—for a distance of between 300 to 1500 feet, beyond which the descent becomes rapid, dropping suddenly to several thousand feet. At a distance of less than three-quarters of a mile from the Island of Clermont Tonnerre, the lead was run out to a depth of 3,600 feet, and yet no bottom was found; at a distance of seven miles a run of 6,000 feet failed to strike bottom. Off the Cardoo

atoll soundings made at a distance of 60 yards from the coast
failed to detect bottom at a depth of 1200 feet, and 500 feet
out from Whitsunday Island no bottom was found by Beechey
at a depth of 1500 feet. Captain Fitzroy found that at a distance of 6,600 feet from the Keeling Island shores the lead
did not strike bottom even after having been run out to a
length of 7,200 feet. Seven miles to the north-west of the Bermuda reef, as has already been seen, the depth of water is
2,100 fathoms, but the coast shallows for a considerable distance around the islands. It would thus appear that the
pitch of the coral island beneath the ocean is at a very steep
angle, sometimes considerably exceeding 45 degrees. Indeed,
there are some grounds for concluding that in the deeper
parts the faces are nearly vertical, rising like gigantic walls
from the oceanic abysses.

In view of the peculiar conditions attending coral growth—
the limitation of depth to 100 or 120 feet—the difficulty of accounting for the occurrence of coral structures in some of the
deepest parts of the sea at once becomes apparent. It had, indeed, been assumed that coral islands merely occupied the summits of submerged volcanoes, and that their distribution over
the deep-sea was simply an indication of the existence, in the
region in question, of an equal number of buried volcanic
peaks or mountain backs. Recent researches have, however,
failed in the majority of cases to detect the presence of such
hypothetical buttresses rising to within a few feet of the surface, but, on the contrary, tend to show that at least in some
instances the actual coral portion of the island descends of
itself hundreds, if not thousands, of feet into the ocean.

The genius of the late Mr. Darwin has furnished a theoretical
explanation of the phenomenon which, even if it cannot be
held to be proved or conclusive, has at least the merit of a
strong probability in its favor, and of being in consonance
with well-determined geological facts and conditions. This
"subsidence" theory, which until recently received the almost
unanimous support of geologists, is based upon the evidence of

extensive terrestrial movements, and presupposes the existence
of numerous land-masses rising from the deepest water. Around
these, under favorable conditions, reef-building and other corals
would flourish in abundance, the submerged cone affording a
suitable base for the development of the coral animal. The
external margin of the coral barrier or buttress, which may be
assumed to grow from a possible depth of 120 feet, owing to the
invigorating action of the beating surf, and an increased food-
supply, would probably rise more rapidly than the inner parts,
whose development would also in a measure be checked by the
out-pouring of detrital sediment. A shelving inwardly-slop-
ing collar or bank, having a land-nucleus in its center, would
thus be produced. In the ring thus forming, whose outer margin, through the breaking and heaping action of the sea, would
be lifted somewhat above the general water-level, we have the
skeleton of the future atoll. We may now distinguish three
elements in its construction: the outer ring or collar of coral,
the central nucleus of land, and the encircling body of water
which separates the two.

If at this stage of its formation we conceive the enclosed
island to undergo a slow and gradual subsidence the following
phenomena may be assumed to present themselves. The outer
border of the reef would slowly but steadily build itself up to the
level of the water, the growth of the coral colony keeping pace
with the gradual sinking of its substratum, provided this be not
too rapid. The parts sinking below the line of 120 feet would die
out, and their future purpose would be merely to afford a base
for the super-structure. The island portion, on the contrary,
would sink deeper and deeper, until eventually it might completely disappear. We would then have an outer barrier and
an inner lagoon, with probably one or more communicating
passages between the latter and the sea cut through the coral
growth. This is the typical atoll.

When a reef is separated by a considerable body of water
from the adjoining land it is termed a "barrier" reef, of which
two distinct types, the "encircling" and the "linear" barrier

reef, are recognized. An encircling barrier reef differs mainly from an atoll in that the assumed subsidence has not been sufficient to completely bury the enclosed island, leaving consequently, no internal sea, but merely a separating channel formed within the coral boundary. By further subsidence, it is conceived, the encircling reef would be converted into an atoll. When a coral boundary extends for a great distance in a more or less linear direction it is termed a linear reef, or "barrier" reef proper. The great barrier reef off the island of New Caledonia extends in a N. W. and S. E. direction for a distance of upwards of 400 miles, and that of the northeastern coast of Australia has a linear extension, with interruptions of more than 1000 miles. In the case of the latter the width of the intervening strait is in many places between 50 and 60 miles, with a depth of water reaching 350 feet. The reef patches, themselves, even in their broader parts, rarely exceed one or two miles in width.

Besides the three forms of coral structure—atolls, encircling and barrier reefs—which have been assumed to give unequivocal evidence of subsidence, there is still a fourth type, that of the so-called "fringing" reef, which has generally been considered to afford proof either of terrestrial stability or of actual elevation. These fringing reefs hug the immediate shore line, and may, indeed, be said to represent the incipient stage or starting point whence the other forms of reefs were developed; by slow subsidence a fringing reef would, on the Darwinian hypothesis, be converted into a barrier reef. Fringing reefs are frequently continued as a series of superimposed terraces above the dry land,—an unequivocal proof of elevation. They rarely, if ever, descend in the water to depths much exceeding 120 feet, and, as might be naturally supposed from their manner of formation, are but rarely associated with the other forms of coral reefs.

Applying the Darwinian hypothesis of subsidence to the phenomena of coral structures generally, we may deduce the following: A region of atolls, encircling and barrier reefs is

primarily a region of subsidence—of subsidence now actually
taking place, or only recently completed; *per contra*, regions
characterized by fringing reefs are regions either of stability or
of slow and gradual upheaval.   The greatest area of indicated
subsidence is that of the Central Pacific, which has been assumed to compass a tract measuring 6000 miles in length and
2000 miles in greatest width.   Commencing at the Paumotu
group, or the Low Archipelago on the south-east, and extending to the Carolines on the north-west, the coral structures dot
at intervals the surface of the sea for a linear distance of 100
degrees of longitude, embracing in this space several hundred
true islands, besides numerous reefs of one form or another.
In the Paumotu group alone there are, according to Dana, not
less than 80 atolls.

The existence of such an enormous subsidence area as is involved in the Darwinian hypothesis is necessarily difficult to
realize, and, indeed, numerous apparently valid objections
seem to interpose themselves to its full acceptance.   It has been
shown that within, or immediately on the border of, the supposed subsiding area there occur local tracts where fringing
reefs take the place of atolls; and, again, others where raised
coral patches or terraces clearly indicate elevation.   The coral
on some of the Hervey and Friendly islands is stated to occur at a height of 300 feet above sea-level; on the island of
Guan, one of the Ladrones, according to Quoy and Gaymard,
the coral rock is in places fully 600 feet above the sea.   In
some island groups, as Hawai, Feejee, etc., coral structures apparently indicative of both depression and elevation occur
interassociated among the different islands constituting those
groups, and the same feature—the interassociation of fringing
and barrier reefs with atolls—has been observed by Semper in
the Pelew Archipelago (West Pacific).   This condition, together
with various concomitant difficulties that lie in the way of the
Darwinian hypothesis, has led to the rejection by many
naturalists—Semper, Guppy, A. Agassiz, Murray, Geikie, and
others—of the subsidence-theory, and to the substitution for

it of a theory of simple coral upgrowth, with structural modifications as depending principally upon currental action and food-supply.

This theory, like its alternative, presupposes as a first necessary condition of coral growth the existence of a submarine basement within the zone of coral life (1–20 fathoms). Upon this, which may be the buried slope or the summit of a volcano, or merely a bank, the coral animal develops and builds to the surface. Where such a sub-structure does not immediately exist, or rather does not extend to the zone within which reef corals are limited, it is claimed that suitable foundations may be obtained through the building up of submarine volcanoes by the deposition on their summits of organic and other sediments. This would explain the apparent anomaly of coral structures rising from depths vastly exceeding the lower boundaries of coral growth, a condition which to Mr. Darwin necessitated the assumption of subsidence. It is well known that throughout the greater mass of the ocean there is a constant rain or down-pouring of organic particles in the form of the calcareous and siliceous tests of Foraminifera, pteropods, diatoms, etc., much of which goes to form the vast accumulation of white mud (Atlantic or Globigerina ooze) which covers the greater part of the oceanic floor. Manifestly, such an accumulation must eventually acquire great thickness. It is more than doubtful, however, if any very considerable thickness of such deposit has been built up during the existing period of coral growth, or that an accumulation of this kind has materially aided in building up the sub-coral buttresses of the deeper seas. The investigations of Mr. Murray, deduced from data obtained by the "Challenger," indicate that a column of oceanic water of 600 feet depth, with a transverse area of one square mile, contains some 16 tons of suspended organic particles; these, if precipitated to the floor of the sea, would make a deposit $\frac{1}{10000}$ inch in thickness. It has thus far been impossible to determine the duration of life of the organisms furnishing the organic particles, mainly Foraminifera, and

consequently there is no direct way of ascertaining in what
period the tests of a given column of water are replenished.
But manifestly, there can be no more rapid accumulation of
the calcareous ooze than there is lime-carbonate suspended in
the sea; and again, the quantity of lime-carbonate so suspended must depend upon the quantity of the formative
material contained in the sea—the quantity of lime carried in
by the rivers, and any residual or surplus quantity that might
be already existing. Now, it would seem from careful observation made on many of the most important rivers of the
globe that the quantity of lime carried out by them into the
sea annually is about one-sixth that of their suspended sediment, which would cover the sea-bottom, if precipitated at a
rate proportional to that of the removal of continental sediment—one foot in 3000 years—to a depth of about $\frac{4}{18000}$ inch.
Assuming that one-half of this amount is used by the Foraminifera for the construction of their shells, the rest being
taken up by the mollusks, corals, etc., then the foraminiferal
accumulation from this source would be the $\frac{2}{9000}$ part of an
inch annually, or very nearly the amount that would accumulate from the droppings contained in the 600-foot column of
water, as deduced from Mr. Murray's determination. At this
extremely slow rate of accumulation, it would require a period
of 100,000 years to build up the thickness of a single foot!
Naturally along coast-lines, where the molluscous animals
largely contribute to the general growing mass, and where inorganic sedimentation is unusually brisk, the process of upgrowth
may be comparatively rapid, especially in the trend of powerful
oceanic currents. A condition of this kind seems to obtain
along the Floridian coast, and it is not unlikely, as has
been suggested by A. Agassiz, that the Florida banks have
been built up largely in the manner above described. But the
conditions become very different when the oceanic abyss, such
as the central Pacific, is substituted for a comparatively shallow
coast-line. Indeed, even in the case of the Floridian banks it
is doubtful if most of their upgrowth is not really due to

bodily uplift rather than to organic and inorganic accumulation, as we have most conclusive evidence of an uplift in the
peninsula of Florida in a period at least as late as the Pliocene. Nor are evidences of a more recent contrary movement
wanting in the same region.

It will, however, naturally be urged against this assumption
of slow accumulation that the quantity of the salts of lime
already contained by the sea is vastly in excess of that which
is annually thrown in by the rivers, and that, therefore, the
amount of formative material on hand is amply sufficient to
meet all the exigencies of a rapid growth. The quantity of
calcium actually contained in every cubic mile of sea-water is
estimated to be nearly 2,000,000 tons, while that held by an
equal volume of river-water is less than 150,000 tons. At the
rate of the present carrying capacity of rivers it is calculated
that it would require 680,000 years to pour into the ocean an
amount of calcium equal to that which is now held by it in
solution.* The question here naturally presents itself: To
what extent is this surplus quantity of lime drawn upon by the
oceanic organisms for the construction of their hard parts or
skeletons? It is in the nature of things impossible to give a
direct answer to this question, but the following considerations
suggest themselves. As far as our knowledge permits us to
pass beyond the region of facts, we can but assume that the
salinity of the sea is progressive or cumulative, and not the
reverse, and that the saline constituents of ocean water are
primarily the products of destruction arising from the wear
and tear of the land-surface. There seems to be no good reason for supposing that the quantity of salts in the sea, and of
lime especially, was ever much in excess of what it is to-day,
unless it was near the beginning of geological time; on the contrary, there are some grounds for concluding that this quantity
may have been less, and even considerably less. If this conception is true, it is manifest that, as far as organic consump-

---

*Murray: "Structure, Origin, and Distribution of Coral Reefs and Islands."
Nature, Feb. 28, 1889, p. 426; 480,000 years according to Reade.

tion of lime is concerned, there is either existing stability in
the sea, or that the different shell-bearing animals remove less
of the formative material for their own purposes than the sea
receives from continental erosion. In the calculation before
made we have used as a basis merely the quantity of lime-carbonate carried out in solution by rivers; to this must necessarily be added that which is derived directly by the sea
through its own breakages—the wear of the coast-line—and
the other salts of lime of which no account has been taken. If
we double the quantity that has been assumed we will probably more than cover the available supply; a rate of accumulation, therefore, of one foot in 50,000 years would be the result.
It is needless to say that such a slow accumulation is hardly
compatible with any notion of growth from great depths, and
that it is entirely opposed to the view which holds to the
formation of giant banks leading up to the zone of coral life.*

But in what, it might be asked, lies the direct evidence that
giant banks are being built up through organic accumulations?
Is it merely the finding of foraminiferal and pteropod ooze on
projecting knobs of the ocean bottom? This is not a new condition, and it is practically repeated in the Globigerina ooze
which covers much of the oceanic floor. It would, indeed, be
remarkable if such deposits did not exist, but their presence
gives no answer to the possibility of building up giant banks
under the conditions which would be considered necessary for
the making of coral islands. No one has more carefully studied,
or is better acquainted with, the Florida reefs than Alexander
Agassiz, and perhaps no class of reefs has been more frequently
appealed to in the recent discussion of coral structures than
those examined by this authority. We are informed by Mr.
Agassiz that these reefs are merely organic growths and ac-

*In evidence of the possible rapid accumulation of a foraminiferal and
pteropod deposit, and the building up of submarine banks, Prof. Hickson (Address before British Assoc., Bath, 1888) instances the case of the basal limestone of the
elevated reefs of the Solomon Islands, to which attention has been called by Guppy.
But manifestly this limestone was formed in shallow water, where the conditions for
rapid organic accumulation are almost infinitely more favorable than they are in deep
water.

cumulations, whose present positions, whether of horizontal or
vertical distribution, have practically no connection with recent movements either of elevation or depression. "There is
practically no evidence that the Florida reef, or any part of the
southern peninsula of Florida which has been formed by corals,
owes its existence to the effect of elevation ; or that the atolls of
this district, such as those of the Marquesas or of the great
Alacran Reef, owe their peculiar structure to subsidence."[1] On
what evidence, it might be asked, rest these assertions ? It may
not be easy to prove subsidence in the case of the Marquesas
and the Alacran Reef, but I believe it would be equally difficult to prove the reverse proposition—*i. e.*, that there has been no
subsidence. As far as the Florida reefs themselves are concerned,
I believe the evidence is all but conclusive that they owe much,
if not most, of their existence to uplift, and to uplift within a
recent geological period. My own researches in the southern
part of the peninsula have demonstrated the existence of Pliocene deposits in vast horizontal, or nearly horizontal, beds as far
south as the Caloosahatchie, and there can be no question that
these deposits, which rise to 10 or 15 feet above the level of the sea,
are continued for some distance still further to the south.[2] The
same deposits, moreover, are capped by deposits of Post-Pliocene
age, proving that an uplift took place in this region as late as the
Post-Pliocene period. That this uplift should not have affected
the apex of the peninsula, and even the reefs beyond, seems
scarcely credible. From what we now know of the structure of
the Floridian peninsula it is clear that this portion of the
North American continent represents a comparatively old
chapter in geological history, and that it has passed through
much the same phases of construction as the border area of the
Eastern and Southern United States. Its periods of elevation
and depression, extending back through the greater portion of
the Tertiary epoch, were largely coincident with those of the

---

[1] "Three Cruises of the Blake." I, p. 61. 1888.

[2] "Explorations on the West Coast of Florida and in the Okeechobee Wilderness," 1887.

regions above indicated, and the movements were with little doubt long sustained, and certainly affected large areas at a time. There is nothing, as far as I can see, to indicate that these movements were confined to what is now dry land; the more natural conclusion is that the axial or plateau uplift extended much beyond the limits of the present peninsula, and as well southward as westward or eastward. The similarity in the geological structure of Yucatan, as it appears from our present knowledge, lends weight to the supposition that the area thus affected by movements was perhaps continuous completely across the Gulf.

In explanation of the distinctive form of atolls—the ring of coral with its inclosed lagoon—it is claimed by the opponents of the subsidence theory that coral plantations building up from submarine banks will grow more rapidly on their outer margins, where the food supply is the greatest, and where, as compared with the inner parts of the mass, there is less obstructive sediment, and thus an exterior rim or elevation would be formed. The differentiation of the inner and outer parts, it is assumed, would be further intensified by the removal in solution of the lime-carbonate from the less active interior portion—the region of coral decay and detrital accumulation—and the formation there of a shallow pan of water or lagoon. That the distinctive features of an atoll may be brought about somewhat in the manner here described can scarcely be doubted; indeed, the supplemental atolls of diminutive size that so frequently accompany the larger reefs, the serpula-reefs of the Bermudas for example, convincingly prove the possibility of ring structure without subsidence. But in instances of this kind the ring is merely a narrow projection, barely rising above the shallow central depression, and is due probably more to the action of a beating surf than to any other cause. In the case of a true atoll with a large lagoon the conditions are very different, and it seems impossible to explain the central depression, often 20, 30, and 40 fathoms, or even more, in depth, on the assumption of internal solution, aided by external acceleration as dependent upon an increased food supply. It does not appear exactly clear why solution should progress more rapidly within the lagoon than over the deeper slopes of the coral buttress, where the protective power of the living animal is also wanting; nor is it at all likely that such solution as actually does take place within the lagoon more than compensates for the accretion of sedimentary material derived from the destruction of the surrounding shores, or for the organic accumulation that is continuously forming along the floor of the lagoon.

My examination of the Bermudas convinces me that, as far as those islands are concerned, the quantity of lime removed from the interior waters is far less than that which is added through sedimentation and organic development. The bottom is everywhere covered with fine debris, and the even floor indicates that this debris is of considerable thickness. One has but to gaze upon the undercut and crumbling ledges of Harrington Sound and the cliffs facing the lagoon to be convinced that accumulation, and not solution, is the prevailing condition in these waters. Yet we have here a depth of water of from 50 to 80 feet. I am, indeed, far from convinced that the organic accumulation which is here taking place by actual growth does not far surpass the material removed through solution. The tests, both perfect and fragmentary, of Foraminifera are abundant everywhere, but in addition to material derived from this source, there exist large areas which are seemingly well covered with the shells of molluscous animals (Chama, Arca, Avicula, etc.) and sea-urchins (*Toxopneustes variegatus*). The latter, with *Arca Noæ*, are especially abundant. The coral growth of Castle Harbor, and not less the insular patches of millepore, etc., in the big lagoon, speak with sufficient emphasis on this point. There can be no doubt, too, that some of the basins and channels have been recently shallowing through silting, but of course this may have been brought about through a mere transference of material from one point to another. The depth of water in the Flatts Inlet, which receives a strong tidal current from the outer lagoon and from Harrington Sound, is much less to-day than it was in the early part of the century, when the Inlet furnished a safe anchorage to vessels of large draught.

Mr. Bourne finds similar conditions to exist in the lagoons of the Diego Garcia reef, and he entirely rejects the theory that lagoons could have been primarily formed through solution. He shows that nowhere has the lagoon deepened since the time when Capt. Moresby surveyed the region in 1837, but, on the contrary, evidences of shoaling to the extent of a full fathom on the south side are not wanting. It is also pointed out that the depth of water in the lagoons of the various islands which are associated with Diego Garcia is not proportional to the size of the lagoon, as we should naturally expect to find it in accordance with the theory of solution. This is also true of the Bermudian waters, although their relations somewhat differ from those of the Chagos Banks. Thus, the depth of water in the comparatively small Harrington Sound is measurably greater than that of the outer water, the big lagoon; it is also much greater than we find it in the superficially more extensive Castle Harbor.

Experiments made to determine the solvent power of seawater show that the process of solution is a very slow one. It appears indeed incredible, in the face of such energetic solution as is presumed to exist in the upper waters of the ocean, that any extensive organic accumulation could ever take place over the floor of the sea, where the solvent power of the water is materially increased through pressure, and still less possible that any considerable foundation could be built up from it, or from the summit of only a moderately depressed mountain peak. The fact that in so large a number of atolls the lagoons are either entirely wanting, or are reduced to mere shallow pans of water, also militates against the hypothesis of solution.

With regard to the formation of the primary ring through accelerated growth on the outer margin, as depending upon

BERMUDA COTTAGE.

an increased food-supply, it may be reasonably doubted if this condition could obtain in the open ocean away from a land area, inasmuch as by far the greater quantity of the food-supply would be given to the polyps as a direct down-pouring from above, and independently, or nearly so, of any currental action. It is true that the outer polyps or colonies would be favored by having an extra supply on their exposed borders, but this would tend probably in the majority of cases only to lateral extension, or to lateral extension combined with upward growth—in other words, to a simple turbinated growth with a nearly flat top. It is true that in a few instances, as has been noted by Semper and Darwin, colonies of Porites, having a turbinated form, exhibit a raised border or lip, but it is equally true that in by far the greater number of cases the individual larger colonies assume either a clavate or a hemispherical form, the latter condition being also distinctive of the giant brain-corals. Mr. Bourne, from his researches on the Diego Garcia reef, also dismisses the notion that food-conveying currents are especially instrumental in shaping the reefs, and he points out that frequently the most elevated side of an atoll is turned away from such currents, and, again, that a large number of coral islands are placed entirely to one side, or out of the path, of the prevailing ocean current.

But even granting that through some method of accelerated growth on the exterior an elevated bounding ring should be formed, the difficulty in accounting for the existence of the deep lagoon would in no wise be lessened; for, in the first place, no such ring would be formed below the line of coral growth, and we should consequently be compelled to assume as antecedent to its formation the complete upward growth or elevation of the submerged bank to the true coral zone, or to a greatest possible depth beneath the surface of 100 or 120 feet. Manifestly, under such conditions there could be no deep depression corresponding to lagoons of 200 or 300 feet depth, unless these were subsequently formed by means other than solution. Furthermore, it appears that the true energy

of coral growth is concentrated in the first zone of some fifty
or sixty feet, which would practically mark the depth at which
a bounding rim of accelerated growth would be formed, and
also fix the depth of the lagoons.*   But as has already been
seen, the depth of nearly all extensive lagoons is very much
greater, in some cases six times as great, or more.

The difficulty in the premises disappears almost entirely if
we accept Mr. Darwin's hypothesis of subsidence, for here the
accelerated outer growth is assumed to depend no less upon
interior retardation (as the result of the accumulation of in-
jurious sediment), as upon an actual increase in the quantity
of the food-supply.   The depth and size of the lagoon will
then depend upon the extent of land that has undergone subsidence, and upon the measure of its submergence.   Where
the descent is very gradual the upward development of the
coral structures may by overgrowth completely close out the
lagoon ; where, on the other hand, the descent is unusually rapid,
more rapid than the compensating upward growth of the corals,
a "drowned" atoll may be the result.   The great Chagos Bank,
which is situated some 700 miles to the south of the Maldives
and has a length of about 90 miles with a greatest width of
70 miles, has generally been assumed to be only a completely
submerged or drowned atoll.   If raised to the surface it would
be in the form of a true atoll, with a depth of water in the
lagoon of 40–50 fathoms.   At the present time the bounding
reef is covered with water of from 4 to 10 fathoms depth.   The
Bermuda Islands have also been instanced as an example of a
partially drowned atoll, but, as has been shown in the preced-

*It is surprising that this consideration in the assumed formation of deep lagoons
through accelerated marginal growth should be so generally overlooked.   Prof. Hickson, in his address before the British Association (1888) on "Theories of Coral Reefs
and Atolls," furnishes an instance of such oversight.   He says: "It seems very
probable then that when a large submarine bank, by accumulation of sediment or by
elevation, comes within the limit of coral growth, the growth commences and is
almost confined to the edges of the bank, and that in course of time the edges of the
bank reach the surface, whilst the centre of the bank has made little or no progress.
This seems to be a very reasonable explanation of the deep lagoons of large atolls,
and one to which at present I can see no valid objection."

ing chapter, there is nothing in the present land-mass to indicate that it bears any direct relation to an atoll ring.

An objection that has been frequently urged against the subsidence theory, and one that has been more particularly insisted upon by Guppy as the result of extended observations made in the Solomon Islands, is that where fringing reefs are exposed they usually exhibit only a moderate thickness of true coral-rock, the basement or sub-structure being mainly of a pelagic character—that is, built up of the remains of pelagic animals (Foraminifera, etc). Hence, it is argued that in the so-called subsidence reefs—atolls and barrier-reefs,—the actual thickness of coral is very limited, or barely more than that which would fall within the regular zone of coral growth. The few observations that have been made on this point, cannot be considered to throw much light upon the question, the more especially as the evidence obtained is far from corroborative. Furthermore, it is just in such elevated reefs that in accordance with the Darwinian theory we should frequently look for a thin deposit of coral-rock, for if there has been elevation instead of subsidence the thickness must necessarily be slight; when, however, subsidence had preceded elevation the result would be the opposite. No weight should be attached to the oft-repeated assertion that in the older geological formations there are no really massive reef-structures. This assertion is entirely opposed to the facts, to cite but a single instance presented by the Dolomites of the Tyrol, the reef-structure of which has been so ably worked out by Mojsisovics and others. Furthermore, it is practically impossible in the case of a large number of the altered limestones to state whether they are of coral origin or not.

One objection against the subsidence theory has still to be considered. It is the association of fringing reefs with atolls. This commingling of two distinct types of structure, implying movements in opposite directions, has been much commented upon, and placed under strong emphasis by the adherents of the new views regarding the formation of coral islands. But the occurrence appears to be entirely without significance.

An alternate movement of elevation and subsidence is no
more strange over an oceanic area than it is on the continental
borders. Yet we have here almost everywhere evidences of a
differential movement, and no geologist has for a moment expressed surprise at the manifestation. What then is the
anomaly of the occurrence of such movements in a coralline sea ?
How is the conception of subsidence antagonized by the facts
of elevation? If we conceive of an atoll, with a deep lagoon,
once having been formed through subsidence, what is to prevent a succeeding elevation from lifting parts of this atoll, or
for that matter, the entire atoll-ring, above the water? We
could still have the lagoon of subsidence retained, and yet as a
last record of movement we would have merely the evidence
of elevation. Because a certain structure is formed through
subsidence it does not follow that this subsidence should not
be followed by elevation. This is but the order of things we
find everywhere expressed in the history of continental masses.
Indeed it would be but natural to look for local oscillations in
regions of extensive movement. Mr. Bourne lays great stress
upon the evidences of elevation (of a few feet) which are presented by Diego Garcia, and claims them to be conclusive
against "the idea of any subsidence being in progress, as Mr.
Darwin fancied to be the case in the Keeling atoll"*. I confess that I can find nothing in this evidence which would preclude an assumption of subsidence sufficiently recent to have
produced the characteristic atoll form. We have in the elevated
beach-rock of the Bermudas unequivocal evidences of elevation,
but equally conclusive are the evidences of the subsidence
which followed this elevation. In other words we have here
the conditions of Diego Garcia simply reversed. Again, in regions where, as in that represented by the great Chagos Bank,
it might be assumed that "drowned" atolls have been formed
as the result of too rapid subsidence, a change of movement
would be all but certain to develop reefs of elevation in combination with those which are assumed to bear in their structure

*Loc. cit., p. 446.

the evidences of subsidence. In other words, there would be an interassociation in the same archipelago of both fringing reefs and atolls, for it can scarcely be conceived that all the projecting land-masses of the archipelago could, at the time when movements of one kind or another set in, have been equally elevated above, or depressed beneath, the surface of the water. Hence, unequal developments must have taken place.

Such are the principal circumstances connected with the history of coral islands. If the theory of subsidence cannot, perhaps, be considered to be absolutely demonstrated, it accords best with the facts, and, indeed, may be said to be in substantial harmony with them. Furthermore, it helps to explain the significant fact, first pointed out by Dana, that a very large, if not the greater, number of coral structures are ranged along the line of greatest depression in the sea.

The question here naturally suggests itself: Is there any evidence supporting the theory of assumed subsidence of the oceanic basins beyond what is furnished by the coral islands? It must be admitted that our positive knowledge on this point is very limited—indeed, almost nothing. But various considerations lead to the belief that the present site of the oceanic basins is a very ancient one, and possibly one that has not materially changed, except in so far as intensification is concerned, since it was first marked out as the most prominent feature of the earth's crust. While manifestly we can have no proof of this condition, it seems but reasonable to assume that if this vast depression was formed through an early flexure of the crust, and as the result of weakness in certain parts of that crust, it has retained its position of depression from the first. With a contracting or moving crust, moreover, particularly under the special conditions of loading (sedimentation) and continental unloading (denudation), it is likely that a depression of this kind would tend to sink or to subside, and force a relief from strain in the uplift of the continents. This is the view now held by probably the greater number of physicists and geologists. But it does not carry with it the assumption of a necessary permanence in the positions of continents and oceans; it does not imply that the oceanic basins were originally of the extent that they are to-day, as we are led to believe by many geologists. It is far more probable that the existing dimensions have been brought about through progressive or cumulative subsidence, which has gradually swept away land-masses that at one time occupied some of the present area of the sea. The long lines of ridges which have been revealed to us by deep-sea soundings, and the placing on these of many of the oceanic islands (volcanic peaks), together with the evidence which the past and present distribution of animal life carries with it, all support this conclusion. It seems, indeed, impossible to account for the existence of oceanic (volcanic) islands, or for the negative islands which rise as prominences from the oceanic floor to within a comparatively short distance of the surface, except on the assumption of subsidence. What is the significance of buttresses like St. Helena, Ascension, the Caroline Islands, or the giant peaks of the Sandwich Islands rising from depths of two or three miles, or more? Can it be assumed that they have been steadily built up volcanically from the ocean floor, four or five miles in height? This is, perhaps, not impossible, but it hardly appears probable. Vulcanism in one form or another doubtless manifests itself over the floor of the ocean, but all indications point to a comparatively limited action in the greater depths. Were submarine eruptions at all numerous, or of that intensity which might be assumed to be necessary for the construction of a giant mountain-peak, we should be probably made aware of their existence in a manner not less emphatic than in the case of subaerial eruptions. It might be assumed that the long intervals at which eruptions take place would prevent special notice of such phenomena, and that, consequently, their effects, even if most momentous, would be placed practically beyond observation. But this is not likely to be the case. When we consider the large number of peaks that in

one form or another come to, or beyond, the surface, and realize how few of them are in a condition of activity, it is difficult to believe that many of these peaks are to-day in a course
of volcanic construction, or that other submarine peaks, scattered between these, are undergoing a similar process of formation. It seems far more natural to assume that these peaks
or islands have been for a long time fully formed, and that
they were formed at a time when their relations to the surrounding sea were more nearly those which govern the positions of by far the greater number of the active volcanoes of
to-day. In other words, they were probably continental or
sub-continental, and their present positions are the indices of
continental subsidence; the vast mass of overflowing water
may have extinguished the fires that at one time supplied the
material for eruption.

The recent discovery of a large number of submarine peaks,
whose existence had not previously been even surmised, rising
to within a comparatively short distance of the surface, seems
to support the general conclusion of subsidence. The soundings of telegraph ships indicate that between the latitude of
Lisbon and the island of Teneriffe there are not less than seven
peaks over which the depth of water varies from not more than
12 to 500 fathoms. From the entire oceanic basin it is claimed
that there are already known about 300 such "submarine
cones, rising from great depths up to within depths of from 500
to 10 fathoms from the surface"*.

Probably the greatest difficulty that lies in the way of the
acceptance of the subsidence theory of coral structures is the
fact that there are not more islands which are in a condition
of semi-formation—i. e., peaks, partially submerged and surrounded by an encircling barrier reef. This is not an insuperable objection, and might be treated by some geologists
in the nature of negative evidence. But the fact is of significance, and must be taken into account for whatever it may be

*Murray: *Nature*, Feb. 28, 1889, p. 423.

worth, in all theories bearing upon the formation of coral islands.

Since the preceding notes were sent to press Alexander Agassiz has published his observations on the "Coral Reefs of the Hawaiian Islands."* This paper, apart from giving detailed descriptions of the reefs of the Sandwich Islands, presents, on the whole, perhaps the clearest statement of views bearing upon the structure of coral islands that has yet been published, but it can scarcely be said that it contributes materially toward the solution of the general problem. Mr. Agassiz asserts himself to be a pronounced opponent of the theory of subsidence, as, indeed, he has always been since he first undertook the very careful survey of the Florida reefs. I think it will be generally admitted, however, that the evidence which is now brought forward is, as far as the substitute theory is concerned, almost wholly negative, while much of it favors the theory of subsidence. Mr. Agassiz assumes certain definite premises or propositions, which are dogmatically stated, but it is difficult to find the exact evidence upon which these premises are based. The special points of evidence which, in the opinion of this authority, render the subsidence theory unnecessary and untenable are practically the same as those which have already been discussed, and consequently they call for but little detailed consideration.

Mr. Agassiz considers it " remarkable that Darwin, who is so strongly opposed to all cataclysmic explanations, should in the case of the coral reefs cling to a theory which is based upon the disappearance of a Pacific continent, and be apparently so unwilling to recognize the agency of more natural and far simpler causes;" and he further expresses himself: " as long as we can in so many districts explain the formation of atolls and of barrier reefs by other causes, fully sufficient to account for the numerous exceptions to the theory of Darwin, which have been observed by so many investigators since the

*Bulletin Mus. Comp. Zoology, XVII, April, 1889.

days of Darwin and Dana, it seems unnecessary to account for
their presence by a gigantic subsidence, of which although we
may not deny it, we can yet have but little positive proof"
(p. 131). These reflections are well so far as they go, but have
the "natural and far simpler causes" underlying the formation of coral reefs, which are to take precedence over the Darwinian hypothesis, been satisfactorily demonstrated? I believe
not, and I further believe with Dana and Von Lendenfeld that
no facts that have yet been brought forward stand in direct op-
position to the theory of subsidence.* Mr. Agassiz assures us
that the "Mosquito Bank, the Yucatan Bank, and the smaller
banks between Honduras and Jamaica, are all proof that
limestone banks are forming at any depth in the sea, or upon
pre-existing telluric folds or peaks, constituting banks upon
which, when they have reached a certain depth, corals will
grow" (p. 133), and a similar condition is considered to under-
lie the formation of the Florida reefs. It has, however, *not*
been shown that these banks have been actually built up in
the manner that has been described, or that any other banks
have been similarly reared from really great depths. The
assertion that the Florida reefs have not been assisted in their
upward growth by elevation (p. 142) is, as far as I can see, not
supported by fact, for we have in the regular horizontal limestone beds of the southern part of the peninsula the most conclusive evidence of elevation even as late as the Pliocene and
Post-Pliocene periods, and there is every reason to believe,
even if the condition cannot be proved, that this upward
movement did not stop short of the coral-forming tract. Nor
does this movement of elevation preclude the possibility of
subsidences having taken place coincidentally in the same region. It appears to me by no means certain that the deep
channel now separating the apex of the peninsula of Florida
from Cuba, and known as the Straits of Florida, was really
cut by the Gulf Stream, as is maintained by Mr. Agassiz. It
seems to me far more probable that it, as well as some of the

* *Naturwissenschaftliche Kundschau*, Oct. 13, 1888.

other deep channels separating the West Indian Islands, was formed through subsidence—the result of localized breakages in the crust. This view has already been expressed by Suess,[1] who draws a close parallel between the physiographic construction of the basin of the Gulf of Mexico and that of the Mediterranean.

Mr. Agassiz thinks it "somewhat surprising that, in the discussion which has lately been carried on in the English reviews by the Duke of Argyll, Huxley, Judd, and others, regarding the new theory of coral reefs, no one should have dwelt upon the fact, that, with the exception of Dana, Jukes, and others who published their results on coral reefs soon after Darwin's theory took the scientific world by storm, not a single recent original investigator of coral reefs has been able to accept this explanation as applicable to the special district which he himself examined" (p. 133). This condition may be surprising, but it is not less surprising that the different investigators who have rejected the Darwinian hypothesis should have thus far failed to agree among themselves as to their own special theories. Thus, the "solution theory" of the formation of the atoll-lagoon, which has been so much emphasized by Mr. Murray, and the possibilities of which we have already discussed, is practicably rejected by Bourne, Guppy, and Wharton[2], and even Agassiz expresses himself not fully satisfied with its efficiency. And as far as I know no satisfactory explanation of the formation of the deep lagoons has been given by any of these investigators. Captain Warton has recently described[3] a number of submerged reef-structures in the China Sea which have a deep flat centre, surrounded by an elevated growing rim; it is assumed that were this rim to grow up to the surface we would have the characteristic features of an atoll, with its deep central lagoon, presented. But

---

[1] *Antlitz der Erde,* I.

[2] *Nature,* Feb. 23, 1888.

[3] *Loc. cit.,* p. 393.

there is no evidence to show that these submerged atoll-like banks are not really banks of subsidence, rather than of upward growth, and in their general features they do not differ from the Chagos Bank which Mr. Darwin considered to represent a half-drowned atoll. Until a satisfactory explanation is furnished of the origin of these central lagoons, so long must any theory bearing upon the formation of coral structures be considered merely tentative. In the case of the Bermuda Islands, which limit the field of my own investigations in this direction, I am confident that, whatever may have been the original construction of the region, the present lagoon features have been brought about through subsidence; and this conclusion was reached before me by Prof. Rice, who seems to have been amply satisfied with the subsidence theory!

On one point in connection with his recent survey Mr. Agassiz furnishes important testimony, and that is as to the actual thickness of the coral-made rock, or, at least, the depth beneath the surface at which this rock occurs. This has been determined by the artesian borings made in the vicinity of Honolulu, and elsewhere. At various points the bore pierced coral-rock at depths of 100–500 feet beneath the sea-level. In the well of Mr. James Campbell, near the Pearl River Lagoon (?), 28 feet of white coral was struck at a depth of nearly 1000 feet below high-water mark (p. 153), and again at "Waimea, Oahu, 900 feet was drilled through hard ringing coral rock" (p. 152). In these facts, however, Mr. Agassiz sees no evidence of subsidence. He prefers to account for the great thickness of the coral rock "by the extension seaward of a growing reef, active only within narrow limits near the surface, which is constantly pushing its way seaward upon the talus formed below the living edge. This talus may be of any thickness, and the older the reef, the greater its height would be, as nothing indicates that in the Hawaiian district there has been any subsidence to account for such a thickness of coral rock in its fringing reef" (p. 154). But where are the evidences which support this explanation? I must confess that I fail to see any. The assumption of a seawardly-extending talus of coral is, it appears to me, purely gratuitous. Indeed, with the very gentle slope that these islands have beneath the sea it is extremely doubtful if any extensive talus could accumulate as a result of either downflow or downwash. Prof. Dana has well supplied the argument on this point, and its seems to me that it is unanswerable. With a gradient of perhaps eight degrees, and not impossibly much less, it is almost inconceivable that there should be much lateral spread of detached coral boulders. Neither wave-action nor the action of the oceanic currents, except possibly under conditions of earthquake disturbance, would be likely to effect the required displacement.

Again, it might be asked, what kind of direct evidence must we look for to establish the point that there has been no great progressive subsidence in the Hawaiian Islands? The needed evidence is just of that kind which it pleases the Earth to keep to herself, and after which the geologist has in most instances sought in vain. The fact that cinder-cones are found "with their base close to the present sea-level" proves, it appears to me, nothing in this connection, and I fail to see the argument which draws from their existence a proof of non-subsidence. But Agassiz himself admits that there is "some evidence of subsidence [about 50 feet] on the southern shore of Hawaii" (p. 154).

On the whole, it seems to me, that the facts as they are presented are, if they indicate anything at all, directly in favor of subsidence, and of subsidence on an extensive scale. They are in my mind far more conclusive than the somewhat similar facts which have been generally accepted by geologists to prove depression or subsidence in delta-deposits, such as those of the Mississippi or Ganges. Dr. W. O. Crosby, in his paper on "The Elevated Coral Reefs of Cuba,"* shows that the coral limestone of Cuba is in places at least a thousand feet in thickness, and he naturally infers that there must have been

---

*Proc. Boston Soc. Nat. History, XXII, 1882–83, p. 124.

subsidence to nearly this amount. Mr. Agassiz, commenting on this important observation, says (p. 150, *note*) that it does "not throw any additional light on Darwin's theory of subsidence; it is of the same character as all the statements which prove the subsidence by the existence of coral reefs, and while there may have been coral reefs formed during subsidence, it does not prove that their growth is due to subsidence any more than the presence of elevated reefs proves them to be due to elevation." This criticism is in a measure valid, but it must be remembered that one of the "strong" points urged by Guppy and others against the subsidence theory was the (supposed) non-existence of massive deposits of coral-limestone, or such as indicated formation through protracted subsidence. But here we surely have such a limestone (provided the observation is correctly made), and its presence removes what might have been a valid argument against the Darwinian hypothesis. And further, there is reason to believe that the thousands of feet of reef-structure which have been described by Sawkins in Jamaica are largely, if not mainly, of coral growth, and represent a formation produced during a long period of subsidence.

In the foregoing discussion of the structure of coral reefs, as also in the chapter treating of the physical history of the Bermudas, I have used the term "subsidence" (and necessarily its opposite—elevation) in a relative sense, indicating a depression or submergence of the land beneath the sea. But whether this submergence was due to a positive movement on the part of the land, or to a change of level (rise) in the water, cannot readily be determined, as the phenomena attending either form of movement would be practically identical. The broad problem of oceanic transgression and continental stability, which has been so forcibly outlined by Suess, cannot be properly treated in this place.

V.

THE RELATIONSHIP OF THE BERMUDIAN FAUNA.

Mr. Wallace, in "Island Life," has ably discussed the more general features of the Bermudian fauna, and analyzed the conditions which gave to the fauna its distinctive characters. The new material which we were fortunate to obtain enables us to enter further into the discussion, and to supplement and expand the conclusions which had been reached from the study of only a limited number of animal groups.

In its broader aspects the Bermudian fauna is strictly non-continental; it lacks those elements which we associate with the animal life of any extended land area, while negatively, in the paucity of animal forms in general, it presents a character-istic of insular faunas. The deficiencies in both the higher and the lower groups of animals are well marked, and the number of special types represented is not very great. The vast body of water which separates these islands from the mainland has, as might have been anticipated, largely prevented the crossing of American animals, and this is true of all the groups except volants. Barring the two species of whale—right-whale and sperm-whale—which visit the waters of the archipelago, the only "wild" mammalian forms of the region are bats, rats, and a possible shrew (Sorex). The animal supposed to be a shrew is referred to by Matthew Jones (Mammals of Bermuda, Bull. U. S. National Museum, 1884), but unfortunately no positive identification has been made. Four species of rat—the brown or Norway rat, the black rat, the tree or roof rat (*Mus tectorum*), and the common mouse (*Mus musculus*)—are recorded ; the black rat, as elsewhere, is rapidly disappearing, and is now on the verge of extinction. There is little reason to doubt that some, and possibly all, of these forms were transported to the islands in the holds of vessels, just as they have been carried from Europe to America, but no absolute date can be fixed for the first rat visitation. If the accounts of Jourdan are to be credited, no rats were known prior to about 1610, although only a few years later (1618) the islands appear to have been largely overrun by the tree-rat, and to such an extent that, as Captain John Smith, in his History of Virginia, says, " there was no island but it was pestered with them ; and some fishes have been taken with rats in their bellies, which they caught in swimming from ile to ile; their nests had almost in every tree, and in most places their burrowes in the ground like conies; they spared not the fruits of the plants, or trees, nor the very plants themselves, but ate them up " (Jones, p. 158). This great abundance, as Matthew Jones well remarks, points to a much earlier colonization than the purely historical data indicate, allowing even for the most rapid development that these animals are capable of. It would, indeed, be somewhat surprising if these animals had not made an earlier appearance, for it can be readily conceived that at least some individuals, more particularly of the tree-rat, would have found their way over, if not through the agency of vessels, on the drift timber which must at times have reached the islands. The narrative of Americus Vespucius, as bearing upon the islands of Fernando de Noronha, is interesting in this connection, since it shows that these islands, which lie directly in the line of the westward-sweeping equatorial currents, were inhabited by a form of big rat as early as 1503, the year of Vespucius's fourth voyage. How and whence this animal came to the islands it is impossible to say, but not unlikely, as has been suggested by Mr. Wallace and Prof. Branner,* it, together with a species of amphisbænian, may have been currentally distributed from Western Africa, a supposition that

*Branner, Fauna of the Islands of Fernando de Noronha—Amer. Naturalist, Oct., 1888, p. 871.

seems not unlikely in view of certain anomalies of distribution which are presented by the Bermudian fauna.

Up to the present time there have been recorded from the Bermudas four species of bats, two of which, the silver-haired bat (*Vesperugo noctivagans*) and the hoary bat (*Atalapha cinerea*), are common North American forms, while the remaining two (*Trachyops cirrhosus*—a vampire—and *Molossus rufus*, var. *obscurus*) are more strictly tropical, ranging over much of South America and the West Indian Islands. Specimens of the last two, coming from the Bermudas, are in the collections of the British Museum, and appear in Mr. Dobson's Catalogue (1878). These animals must, however, be of extremely rare occurrence in the islands, since they were unknown to both Mr. Jones and Mr. J. L. Hurdis, the latter a fourteen years resident. The silver-haired bat is about equally rare, as but a single specimen seems to have been noticed in the island group, and that one nearly forty years ago.[*] Even the commoner form (*Molossus obscurus*) is very uncommon, and appears only in the autumn months, when the westerly storms bring over numbers of American birds. So rare, indeed, are these animals generally that they are seemingly unknown to the majority of the inhabitants, and even among the older residents I found but little knowledge of cheiropterology. It is singular, in view of the scarcity of these animals, that Capt. Nelson should have considered the "red earth" of the Bermudas, and also of the Bahamas, to have been formed largely as an accumulation of the rejectamenta of bats, which inhabited "once-existing caverns" (Journ. Geol. Soc. London, IX, p. 209). We observed no bats in any of the caves or caverns which we visited, nor did "ancient" guardians of these caverns know anything about such animals. The very rare occurrence of bats in the islands, and the circumstance that they are most conspicuous during the periods of heavy storms, prove almost conclusively that these animals are merely in-blown stragglers.

---

[*] Jones, Mammals of Bermuda, *op. cit.*, p. 145.

COCOANUT PALMS

The bird fauna of the Bermudas, including both land and water forms, comprises, as far as is known, some 187 or 188 species, which, with two or three exceptions, are members of the North American fauna. These exceptions are the sky-lark (*Alauda arvensis*), common European snipe (*Gallinago media*), and gold-finch (*Carduelis elegans*). The first, of which but a single specimen has been obtained, has generally been considered to be an escaped cage-bird, but Savile Reid, in his review of the birds of Bermuda (Bull. U. S. National Museum, No. 25, 1884, p. 178), believes it to have been more likely an inblown straggler. Seemingly only two specimens of the European snipe have been recorded, both of them from Pembroke Marsh, where they were shot in December 1847, by Colonel Wedderburn. The single specimen of gold-finch was observed by Savile Reid near Harrington Sound, in April, 1875; it was very wild, but is still supposed to have been an escaped prisoner. Two other European birds, the wheat-ear (*Saxicola œnanthe*) and land-rail (*Crex pratensis*), have also been noted in the Bermudas, but both of these find their way to Greenland and the mainland of America, so that their occurrence is less remarkable than that of the other forms.

Of the entire Bermudian avifauna somewhat less than one-half the species are land-birds, and of these a fair proportion have been observed only on one or two occasions. There appear to be but eleven permanent residents, nine land-birds, and two water-birds, to wit: cat-bird (*Galeoscoptes Carolinensis*), blue-bird (*Sialia sialis*), white-eyed vireo (*Vireo Noveboracensis*), English sparrow (*Passer domesticus*), cardinal-bird (*Cardinalis cardinalis*), crow (*Corvus Americanus*), Virginia quail (*Colinus Virginianus*), ground dove (*Columbigallina passerina*), great blue heron (*Ardea herodias*), Florida gallinule (*Gallinula galeata*), and tropic-bird (*Phaëton flavirostris*). Two or three species of shearwater (*Puffinus Anglorum, P. obscurus, ? P. opisthomelas*) have at intervals been found breeding in the Bermudas, but seemingly they have now deserted these islands for other quarters.* Mr. Savile Reid informs us that the presence of the Virginia quail or "bob-white" marks a recent introduction, the bird having entirely disappeared from the islands with the year 1840; an importation from the United States was made in 1858 or 1859, and is the orgin of the existing stock of birds.

Mr. Witmer Stone, one of my assistants, has furnished me with the following notes on non-resident birds observed by him during our visits to the islands in the month of July, a season of the year when the bird fauna is probably at its minimum :

Wilson's petrel (*Oceanites oceanicus*). A single individual seen in the wake of the steamer a short distance out from the islands.

Least sandpiper (*Tringa minutilla*). Several individuals seen near Spittal Pond, July 15.

Piping plover (*Ægialitis meloda*). A single individual, which followed in the wake of the departing steamer for the better part of a day.

Green heron (*Ardea virescens*). A single individual observed in the mangroves of Walsingham.

We heard or saw all the resident birds of the islands with the exception of the great blue heron (*Ardea herodias*). The first tropic bird was seen before the land was yet sighted, and from this time until our departure we seldom lost sight of these beautiful creatures. At the time of our visit the breeding season was nearly over, and the nearly fledged young were to be seen sitting on the ledges overhanging the waters. The single egg is deposited in holes in the rock, which are apparently excavated by the parent. We found the birds breeding both on Harrington Sound and on the south shore. The little English sparrow was also found breeding among the shelving rocks of Harrington Sound. We observed but three crows during our sojourn, and it would appear that this bird, which was at various times abundant, even as early as the beginning of the

*Mr. Wallace mentions the coot as a permanent resident, but probably the bird intended is the gallinule.

seventeenth century, and as late as the last decade, is again becoming rare.

The most regular and abundant, among land-birds, of the "regular" visitants are the small-billed water-thrush (*Seiurus Noreboracensis*), snow bunting (*Plectrophanes nivalis*), bobolink (*Dolichonyx oryzivorus*), night hawk (*Chordeiles Virginianus*), and belted kingfisher (*Ceryle alcyon*), some of which arrive and go with almost strict punctuality to season. The pigeon-hawk (*Hypotriorchis columbarius*) and osprey (*Pandion haliaëtus*), as also one or two species of owl, are somewhat less regular, but not exactly uncommon. If we except the Seiurus, all these birds are either partial migrants or hard fliers, and it is not difficult to account for their presence in the islands. Some of them, doubtless, reach the Bermudas in the direct line of their migration, and are not wind-borne. The regularity of the arrivals proves this almost beyond question, as it likewise does in the case of the numerous water-fowl—sandpipers, plovers, snipe, etc,—which so largely abound during the seasons of migration. The condition is otherwise with the birds that have been met with only at long intervals or on single occasions. There can be no question that these are wind-swept, and have been involuntarily carried seaward by sudden storms. Some of the more delicate birds, such as the warblers, tits, and humming-bird, have thus managed to reach the islands, while, doubtless, many more perished in the interval separating them from the mainland. It is interesting to note that even such a large bird as the American swan should have crossed this stretch of the ocean, but it is difficult to conceive that the presence of this bird is due to simple wind-drift. May it not be a case of misdirected flight, following the lead of some other birds? Possibly the exceptional occurrence of the flamingo (*Phœnicopterus ruber*) may be similarly accounted for.

Excluding the marine turtles which visit the waters there is but a single reptile in the islands. It is a skink, *Eumeces longirostris*, a form closely related to the common skink of the

Eastern and Southern United States, *Eumeces fasciatus*. The animal is said to be very common, but we saw and obtained but a single specimen. Until within the last few years the batrachians were wholly wanting from the Bermudas, or at least supposed to be so. Latterly, specimens of the big *Bufo marinus* were introduced, and seemingly the new toad does well. We saw several of these animals in the brackish waters near the Devonshire marshes. Not unlikely a species of coecilian also belongs to the Bermudian fauna, and may indeed be indigenous. We obtained from under a stone a number of eggs beaded to one another in the form of a string, which I was unable to place. Prof. Ryder, of the University of Pennslyvania, has kindly examined these for me, and he believes that they are the eggs of coecilians. They certainly bear a very close resemblance to the figures and descriptions of the ova of the Coecilia, and most so, perhaps, to those of the genus Coecilia itself. It would be interesting to determine to what animal the eggs really belonged. They measured about 5 mm. in diameter.

The preceding enumeration of species brings prominently to light three important points in zoogeography : 1. The impoverished character of the vertebrate fauna ; 2, the distinctively American, and more particularly, North American, aspect of this fauna ; and 3, the general absence of forms peculiar to the islands. These conditions would seem to imply a permanent (past) isolation of the islands from the nearest mainland, and a comparatively brief existence. But this need not necessarily have been the case. Even if we admit a former connection with, or near approach to, the American continent, for which, however, there appears to be but little, if any, satisfactory evidence, we could scarcely hope, under the conditions which have marked the history of the Bermudas, to have retained many elements of a continental vertebrate fauna. The restricted area and absence of freshwater, in conjunction with the depredations of birds of prey, would have soon exterminated, or all but exterminated, what there may have been of mammals, reptiles, and amphibians, leaving but a few groups—bats, rodents—as possible survivals. Nor could we reasonably expect to find the remains of such animals preserved either in the coral rock or in the drift-rock of the islands. With regard to the question of the recent origination of the islands the evidence from the vertebrate fauna proves little. The great distance of the islands from the mainland in itself explains the poverty of the fauna, whether this be old or new, while the absence of distinctive or special types among birds is, as Mr. Wallace well holds, due to the too frequent crossing of migrants or involuntary wanderers, which keeps the various breeds true, and prevents specific modification. The absence of peculiar species is, therefore, not a result of newness; on the contrary, certain considerations seem to indicate that the island group is of greater antiquity than has been generally assumed, and not impossibly some of the lower forms of life now inhabiting it are descendants of an ancient fauna which was well developed before the present physical conditions were established.

An analysis of the Bermudian invertebrate fauna shows some very interesting and remarkable features, which prove the complexity of zoogeographical inquiry. Before our visit but little systematic work—for most of which we are indebted to Matthew Jones—was done in this department of zoology, and, doubtless, much still remains to be done. But the general features of this fauna are now sufficiently determined to permit of satisfactory conclusions being drawn from them.

The marine Mollusca of the archipelago, which up to the time of our visit were listed at about 80 species, comprise, as far as is now known, some 170 species. These, with probably less than a dozen exceptions, are all members of the West Indian or Floridian faunas. Lying in the path of the Gulf Stream drift, which strands upon the island-shores vast quantities of the Gulf-weed (*Sargassum bacciferum*), there seems little reason to doubt that by far the greater number of these forms were given to the region from the south. The species peculiar to the Bermudas are, as far as is now known, about eleven in number, none of which had been described previous to our exploration. They are:

Octopus chromatus,
Aplysia æquorea,
Chromodoris zebra,
Onchidium (Onchidiella) trans-Atlanticum,
Emarginula dentigera,
Emarginula pileum,
Cæcum termes,
Macoma eborea,
Mysia pellucida,
Cytherea Penistoni,
Chama Bermudensis.

In addition to the above there are several shells (Phos, Columbella, Pleurotoma) which I have been unable to place, and which may prove distinctive of this fauna. On the other hand, it is not impossible that some of the forms above named may be found elsewhere, and thus lessen the amount of individuality which the Bermudian fauna now presents. Thus, the *Murex nucceus,* of Mörch, which was supposed to be peculiar to the Bermudas, has recently been found at Marco, west coast of the peninsula of Florida; and a species of Cythara, (unnamed) which I obtained at Shelly Bay, I have since found among unidentified material, from Florida, contained in the collections of the Academy of Natural Sciences. But the facts as they stand are sufficiently suggestive, and in conjunction with much more marked peculiarities presented by the terrestrial Mollusca, point strongly, though by no means conclusively, to a faunal individuality that could have arisen only as the result of long existence, or of a faunal modification that was unusually rapid in its development. How rapid this modification may have been, or how old the islands may be, it is impossible to say, and the terms can, therefore, only be used in a relative sense; but with either condition an antiquity is indicated which extends probably far beyond the time that is generally associated with the making of "recent" coral islands.

The number of terrestrial mollusks credited to the Bermudas is usually given at 19 or 20, but the list which appears further on shows that this number must be increased to 30. An analysis of this list indicates that of the thirty species sixteen, or somewhat over half, are known also from the West Indian islands, five (*Helix vortex, H. microdonta, H. pulchella?, H. appressa,* and *Pupoides fallax*) occur in the United States, three in Europe or the East Atlantic islands (Madeira, Azores—*Helix ventricosa, Helix pulchella, Cæcilianella acicula*), while not less than eight, including all the species of the remarkable group *Pœcilozonites,* appear to be confined to the Bermudas. These species are: *Helix discrepans, Pœcilozonites Bermudensis, P. circumfirmata, P. Reiniana, Succinea Bermudensis, Alexia Bermudensis, Melampus Redfieldi,* and *Helicina convexa.* To this number may perhaps also be added the somewhat doubtful *Helix hypolepta.* The large proportion of special forms, taken in conjunction with the development of a distinct group, is certainly remarkable in the case of an island group which has been generally considered to be recent in formation, but this specialization is also well marked in some of the other animal groups. The fact argues for considerable antiquity, and it is interesting to note in this connection that the ancestral type of the peculiar molluscan genus Pœcilozonites is represented in the common sub-fossil *P.* (*Helix*) *Nelsoni,* the probable progenitor of the recent *P.* (*Helix*) *Bermudensis.*

The conditions governing the dispersal of the terrestrial Mollusca have been fully discussed by Mr. Darwin and Mr. Wallace, and there can be little or no doubt that their explanation of the oceanic transport of these animals is the true one. The floating material of the Gulf-drift has in this instance, doubtless, sufficed to bring most, if not all, of the non-peculiar species of the islands from the West Indies and the Southern United States. A few of the species, again, may have been

transported by vessels, or even through the agency of birds.
The interesting question here naturally presents itself: Of
what relation to the Bermudian fauna are the two or three
identical species—*Helix ventricosa, H. pulchella,* and *Cæcilianella
acicula*—which occur in the East Atlantic islands (Azores,
Madeira, Canaries) and Europe? Are they a part of the western fauna which has gradually drifted eastward, and stocked the
European continent from a home originally insular? The
species (two, at least) have seemingly not yet been found in the
western hemisphere outside of the Bermudas, and possibly they
do not occur elsewhere. If this is the case it is hardly likely
that they could have been carried (except through the agency of
man) from the Bermudas to the Azores or the Canaries, since
the first-named islands lie considerably to the eastward of the
Gulf current, although still within the influence of the Gulf-
drift. The fact that none of the species of the peculiar genus
Pœcilozonites are found in the Azores or the Canaries is a further argument against an assumed eastwardly transport. On the
other hand, it is just possible that the Bermudas have received
these species from the Azores and Canaries through the return
Lusitanian and equatorial currents, and that the Azores furnished to Europe the continental representatives of the species.
There would be nothing strange in this, and the northern position of the return currental flow might explain the absence of
these forms from the West Indian Islands. That islands, which
are favorably situated as far as winds and currents are concerned, should have transmitted to continental areas portions
of their faunas is what we should but expect. It is not only
that the continents furnish the islands, but necessarily the
islands must furnish the continents, but to what extent this
reciprocal action takes place cannot well be determined. From
various considerations Morelet has argued that some of the
molluscan forms of meridional Europe must have originated
in, or, at least, been derived from, the Azores. If the ocean
currents which now pass off the Southeastern United States
trended in the opposite direction there can be no question that

some of the peculiar land-snails of the Bermudas would be drifted to our shores, where, with a favorable climate and vegetable growth, they would soon multiply and spread, and to such an extent as to make it appear as though they originated on the continent.

It might be objected that these seeming anomalies of distribution can be readily accounted for by assuming that there has been simple artificial transport by means of vessels. And, no doubt, full allowance must be made for this contingent distribution. But, again, on this assumption the absence of the commonest species of land mollusks—those which have been most broadly distributed over the earth's surface, and which would have found congenial conditions of environment in the Bermudas—becomes very striking, and equally so whether we consider the forms that may have been transported from the Old World or from the New.

The marine molluscan fauna of the Bermudas is, as has already been seen, overwhelmingly Antillean in character, and there can be no question that its own history is intimately bound up with the history of the fauna of the West Indies. The practically total absence of species of the Eastern United States which are not found in the Floridian waters is astonishing, and shows how insuperable is the barrier which the waters of the Atlantic, and of the Gulf stream particularly, offer to a free migration or dispersion of the species. This, again, appears the more remarkable in the light of certain anomalies of distribution which a critical examination of the species reveals, and which had already in many cases been noted as a characteristic of the West Indian fauna. Thus, the various species of Triton, *Triton chlorostoma*, *T. tuberosum*, *T. cynocephalus* and *T. pileare*, are all members of the fauna of the Pacific and Indian oceans; *Ranella cruentata* crops up in the variety *R. rhodostoma*, from Mauritius. Again, *Epidromus concinnus*, from the Philippines, is represented in our collection by a number of individuals which are absolutely undistinguishable, both in shell ornamentation and color-markings, from the Pacific specimens,

while they differ somewhat from the closely related *Epidromus
Swifti*, from Antigua. A seemingly undescribed form of Columbella (Anachis) is, so far as I have been able to determine,
most nearly related to a species from New Caledonia, *Anachis
plicaria ;* *Natica Marochinensis* is a member of the faunas of both
Western Africa and the Pacific, and *Natica lactea* is apparently
undistinguishable from *N. Flamingiana*, from the Viti (Feejee)
Islands, Philippines, etc.,whence we have also the *Arca imbricata ·*
A number of forms common to the west coast of Africa and to
the southern waters of Europe also occur, but these appear to
be less numerous than the forms which occur in the Pacific
and Indian oceans.   Seemingly but few of the Bermudian
species are found in the Azores (*Purpura hæmastoma, Neritina
viridis, Avicula Atlantica, Pinna rudis*), a somewhat surprising
circumstance in view of the large representation of Pacific forms,
and considering that the Azores lie directly in the path of the
heated waters of the Gulf Stream.   It is, indeed, difficult to
account for these anomalies of distribution, and for still more
marked ones, as we shall presently see, which are presented by
the Crustacea.

   Of molluscan forms which have been hitherto considered
to be restricted to the west coast of America, I can state the
positive occurrence of only two or three species—*Chama
exogyra, Tellina Gouldii*, from the Californian coast.   In the
case of both of these forms I have very carefully satisfied myself as to absolute identity. *Arca solida*, from the west coast,
does not appear to differ measurably from *Arca Adamsi*,
a West Indian form which has its representative in the Bermudian fauna.   I feel satisfied that many more forms are
common to the east and west coasts of America than is
generally assumed to be the case.

   There appears to have been no systematic determination of
the Bermudian Crustacea prior to our visit to the islands.
The collections made by us are not extensive, but probably a
full half of the species which they contain are now for the
first time recorded from the archipelago.   By far the greater

number of species—indeed, nearly all of them—are, as would be naturally expected, forms which belong to tropical or sub-tropical America (Florida, West Indies, Brazil). None of the species, as far as they have been determined—the Isopoda and Amphipoda still await examination—are peculiar to the Bermudas, excepting possibly *Scyllarus sculptus*. The specimen figured by Lamarck in the *Encyclopédie*, and subsequently described by Milne-Edwards, seems to have been "without a home," nor have I been able to trace the species from the writings of later authors. I am, therefore, not in a position to say whether the species is strictly Bermudian or not.

The remarkable fact connected with the Bermudian Crustacea is the appearance of three species of Macrurans which had hitherto been recorded only from the Pacific. These are *Palæmonella tenuipes*, described by Dana from the Sooloo Sea, *Palæmon affinis*, and *Penæus velutinus*, the last a species also first described by Dana. It is remarkable that the only species of Palæmonella other than *P. tenuipes* is likewise an inhabitant of the Sooloo Sea. I am wholly at a loss how to account for the occurrence of these Pacific types at the Bermudas; they may yet be discovered in some intermediate region, and thereby lessen the difficulty in the problem, but for the time being their presence must be considered a zoogeographical knot to be cut. The absence of the common *Palæmon vulgaris*, as well as of the principal crustaceans of the Eastern United States, excepting the more southerly forms, is strikingly noticeable. *Alpheus avarus* is a Eurafrican form; *Pachygrapsus transversus* has been noted also from Australia.

The insect fauna (including here also the spiders) of the Bermudas is distinguished more by negative features than by positive ones; it is eminently deficient. It is not yet known in its full details, but sufficiently so to show that it is mainly a combination of Neotropical (West Indian and South American) and Holarctic (North American) elements. And here we are presented with the significant fact that the insects proper

or fliers are essentially forms common to temperate United
States, while the non-fliers or arachnids are more nearly tropical or sub-tropical forms, or such as have required drift
material to transport them to their present habitation. The
former have evidently been carried across the interposing arm
of the sea in the manner of the rarer birds—*i.e.*, through the
instrumentality of storms. Dr. P. R. Uhler, who has kindly
looked over our collections of Orthoptera, Neuroptera, etc., informs us that strong winds blowing off the mainland of Maryland and Virginia carry countless numbers of nearly all kinds
of insects out over the ocean, and that those that are dropped
into the sea are "returned to the shores by the tides and piled
up in windrows along the beaches. Among these we have
often found the half-drowned dragon flies mixed in with the
thick piles of beetles, bugs, wasps, and flies which stretched
along the line of the retreating tide." That many thus wind-
swept reach to distances at least as remote as the Bermudas
there can be no question.

We observed during our brief sojourn but four butterflies,
Danais archippus, *Pyramis Atalanta*, *P. cardui*, and *Junonia cœnia*,
all common forms of the United States, and these appear to be
nearly all the forms that are usually met with in the islands.
Three or four other species of day-fliers have been observed at
different times, but they are of such rare occurrence as to
barely constitute true elements of the Bermudian fauna. Our
beetles were limited to some five or six species, which Dr. Horn
has kindly determined for me to be *Ligyrus tumulosus*, *L. gib-
bosus*, *Agonoderus lineola*, *Cicindela tortuosa*, and *Opatrinus
anthracinus*, forms common to Cuba and the Southern United
States. A number of other species of Coleoptera have been
collected in the islands, but I am not aware that they have as
yet been carefully determined.

From the list of Orthoptera, Pseudoneuroptera, and Dermop-
tera which Dr. Uhler has prepared for me it will be seen that
they represent types which are included in the United States
fauna of the region between Cape Cod and Florida. Dr.

Uhler calls attention to the significant fact that two of the Pseudoneuropters, *Mesothemis longipennis* and *Lestes unguiculata*, are freshwater types, whose larval condition is dependent upon the existence of fresh, or but mildly brackish, waters. The Bermudian earwig (*Labidura riparia*—*Forficula gigantea*) is a species recently introduced into the Eastern United States from the Mediterranean region.

Prior to 1888 there were but six species of spiders recorded from the Bermudas, of which three were described as peculiar by Blackwall—*Salticus diversus*, *Xysticus pallidus*, and *Epeira gracilipes*. To this number we now add eleven additional forms, one of which, *Lycosa Atlantica* of Marx, proves to be new. Dr. Marx, of Washington, has kindly determined all of our forms, and his notes on species appear on another page. The thirteen species which are not peculiar to the Bermudas are the following:

Loxosceles rufescens, .	W. Indies, Florida, Europe, Asia, Africa, Madeira, Canaries, Cape Verde Islands.
Heteropoda venatoria, .	Cosmopolitan.
Filistata depressa, .	Southern United States.
Uloborus Zosis, . .	W. Indies, Florida, S. Amer., Africa, Asia, S. Helena.
Nephila clavipes, . .	S. United States, S. and C. America.
Epeira caudata, . .	United States.
Epeira labyrinthea, .	N. and S. America (to Magellan), W. Indies.
Theridium tepidariorum,	America, Europe, Azores, S. Helena.
Argyrodes nephilæ, .	United States, Guiana, Peru.
Pholcus tipuloides, .	Samoa.
Dysdera crocata, . .	U. S., Europe, Azores, Canaries, S. Helena.
Menemerus Paykullii, .	Cosmopolitan?
Menemerus melanognathus,	America, Europe, Africa, Canaries, Cape Verde I., S. Helena.

The remarkably broad and somewhat indiscriminate distribution of most of these species shows almost beyond doubt that they have been principally or largely transported through the agencies of commerce. They, therefore, throw but little light upon the subject of zoogeography, although it is interesting to find that such a large number of forms can so readily accommodate themselves to the varied conditions of climate, and of the surroundings generally, which the different countries present. The proportion of peculiar forms is greater than we should have expected to find in a region which is in such frequent communication with the mainland, and is supposed to be of comparatively recent origin. But, as has already been seen, there are good grounds for believing that the islands are more ancient than they are generally considered to be—or, at least, that their fauna is. Mr. Bollman has determined four species of myriapods in our collections, one of which, a *Spirobolus*, is apparently peculiar to the islands. Of the remaining forms, as far as it has been possible to determine from imperfect specimens, one of the species is from the Azores, another from Europe, and the third from the United States. Of course the number of species collected is not sufficiently great to give positive values in the matter of distribution.

Of the lower groups of animals, such as the sponges, corals, and echinoderms, we have principally Antillean and Floridian types represented. That this should be the case, more particularly with the reef-building corals, stands to reason. It is less easy to account for the large number of peculiar or new forms among the holothurians, unless it be on the assumption of antiquity. But they may yet be discovered elsewhere, in the West Indies, although if they existed in the Bahamas, where we should naturally expect to look for them, they could scarcely have failed to attract the eyes of the different naturalists who have from time to time visited the islands. They are in the Bermudas about the most conspicuous objects on the coral sands.

In summarizing the general features of the Bermudian fauna as they have been passed in review in the preceding pages the following broad conclusions and facts present themselves:

1. The Bermudian fauna is essentially a wind-drift and current-drift fauna, whose elements have been received in principal part from the United States and the West Indies. The aquatic animals are overwhelmingly Antillean in character, while the animals of the air—birds and insects—are as overwhelmingly North American.

2. Some portion of the fauna appears to have been derived (through the agency of the return Atlantic current) from the west coast of Eurafrica (including the African Islands), or even from the Azores, while probably but few forms, if any, were given to those regions by the Bermudas.

3. The large proportion of peculiar forms among the terrestrial Mollusca more particularly, and somewhat less so among the arachnids and echinoderms, renders it probable that this fauna is in part of considerable antiquity, and that some of its elements have been developed from a fauna pre-existent in the region when the present physical conditions had not yet been established. This conclusion is supported by the fact that the predecessor of a group of Pulmonata now peculiar to the islands is found fossil or sub-fossil in the rock of these islands.

4. Certain marked elements of the Bermudian fauna are of a distinctively Pacific type—Mollusca, Crustacea—but it seems impossible at the present time to explain this mixed relationship.

5. The currental water which separates the United States from the Bermudas proves a practically insuperable barrier to the direct passage of marine animals from the one region to

the other; hence, the forms of the Eastern United States, except in so far as they may be also members of the southern fauna, are almost entirely absent from the Bermudas.

6. Most of the temperate-American element of the Bermudian fauna owes its establishment on the islands to accidental causes—storm-winds; the tropical (Neotropical element) is, on the other hand, the expression of slow but steady diffusion.

7. Neotropical elements largely preponderate in the permanent or resident fauna of the islands.

8. An arm of the sea may be as insuperable a barrier to the passage of marine animals as it would be to the animals of the land; caution is hence necessary in the discussion of continental and oceanic changes or stability as affecting animal distribution.

THE SOUTH SHORE.

VI.

ZOOLOGY OF THE BERMUDAS.

The following notes on the zoology of the Bermudas are based on personal observations, and on collections made during a brief sojourn on the islands during the past summer, in company with a class of students from the Academy of Natural Sciences. But little systematic work, other than that in the departments of ornithology, ichthyology, and botany, had hitherto been done in this remarkably interesting, and typically oceanic, island group, and it was thought that a more critical survey might bring out facts of general interest to the zoological student, and throw some additional light upon the intricate subject of zoogeography. In the results obtained I have not been disappointed. The exuberance of animal life has yielded much that has proved to be new to the systematist, while certain remarkable peculiarities in the distribution of a number of well-known types of animals open up vistas in geographical distribution which appear to me at present to recede into darkness, and, perhaps, tend to draw only more closely the veil over this mysterious subject.

The specimens noted or described in the following pages were largely obtained through dredgings, which were carried on as well in the outer water as in the smaller interior sounds and lagoons. As might have been anticipated the greatest profusion of animal life was found on the edge of the growing reef itself, the shoals surrounding the cluster of rocks on the northern barrier known as the North Rock. The wealth of forms occurring here almost transcends belief; unfortunately, the limited time at our command and the state of the weather prevented more than a cursory examination of this locality, which is made comfortable for collecting and wading during a partial exposure above water of some three hours. All the dredgings were confined to depths within 16 fathoms, which also represents the greatest sounding made by us in the lagoons.

ACTINOZOA.

The true stone corals of the Bermudas are comprised, so far as we now know, in some twenty-five species, the greater number of which are represented by identical forms in the Bahaman or West Indian seas. The genera thus far indicated are *Oculina*, *Mycedium*, *Astræa*, *Siderastræa*, *Porites*, *Isophyllia*, *Maandrina*, and *Diploria*. The genus *Madrepora*, one of the commonest of the Bahaman and Floridian corals, appears to be absent. On the south and east side of the island group the outer margin of the growing reef, largely covered by a serpuline and vermetus growth, approaches to within a few hundred feet of the shore, where it breaks the inflowing surf into a white crest. Within the line of these breakers the depth of water is in places as much as ten or twelve fathoms. The brain coral (*Diploria*) and various gorgonians develop here in great profusion, the huge yellow masses of the former appearing almost everywhere at depths of from ten to twenty feet. Vast growths of millepore also cover the shallower bottoms, presenting in the ensemble a wonderful garden of animal development. This profusion of coral growth is, however, surpassed on the north side, where the reef recedes to a distance of some eight or nine miles from the island-shores, enclosing an extensive body of water whose depth is in general about eight or ten fathoms, and more rarely twelve fathoms. Much the same coral growth is indicated here as on the south side, the large brain-corals preponderating by their masses. While, probably, the greatest profusion of animal life is really met with on the actual edge of the growing reef, this does not appear to be the case with the corals themselves. At any rate, I was unable to satisfy myself that there was any marked difference to be observed between the marginal growth and that which extends gradually backward from the margin into deep water. Indeed, as far as the brain-corals themselves are concerned, it appeared to me that their largest masses were to be found some distance within the bounding reef, and consequently beyond the breaking action of the surf. This condition is again shown in the comparatively quiet and sheltered waters of Castle Harbor, where portions of the platform-bottom may be said to constitute one almost connected mosaic of huge Diplorias. In so far, therefore, the Bermudas differ from the greater number of coral islands, in which, as is commonly stated, there is a marked deficiency in the coral growth within the bounding area, and an equally marked luxuriance on the crest and outer slope of the reef.

In most places the largest corals do not come nearer than a foot or two feet of the surface of the water, the massive brain-corals rarely appearing in water of less depth than five or six feet. But in the shallows off the North Rock we found *Porites astræoides* almost at the surface in low water, and just off the entrance to Harrington Sound, on the north shore, *Siderastræa galaxea* was covered by only about two inches of water. The borders of Harrington Sound are largely overgrown with species of *Isophyllia*, which likewise approach to within a short distance of the surface. In the greater depths of the Sound we found only Oculina, down to ten fathoms, the dredge-net being frequently caught and reversed by their ramose stems; beyond ten fathoms the dredge usually came up empty.

The following species were obtained by us:

Mycedium fragile, Dana.

Two specimens. North Rock?

Oculina diffusa, Lamk.

Harrington Sound.

Oculina varicosa, Lesueur.

Harrington Sound.

Oculina pallens, Ehrenberg.

Harrington Sound.

I feel satisfied that this species is identical with the preceding, the same stock bearing what might be considered to be typical representatives of both forms.

The amount of variation in the disposition of the calyces, as well as in their individual shape, is very great in this genus, and I am by no means sure that two or three of the other forms of Oculina here enumerated represent anything more than varietal modifications. Pourtalès, in his illustrations of the corals of the Florida reefs (Mem. Mus. Comp. Zoology, VII, plates I and II) correctly refers, it seems to me, both types to a single species (*O. varicosa*).

Oculina speciosa, Edwards and Haime.

Harrington Sound.

Oculina recta, Quelch.

One specimen, from Harrington Sound, which agrees in the characters of the species from St. Thomas (Quelch, Challenger Reports, Zoology, XVI, p. 51). The species does not appear to have been hitherto observed in the Bermudian waters.

Oculina coronalis, Quelch.

Harrington Sound. First described from the Bermudas (Challenger Reports, Zoology, XVI, p. 49.)

Quelch, in his report on the reef-building corals of the Challenger (*op. cit.*, pp. 9 and 49), enumerates as an additional member of the Bermudian fauna the *Oculina Bermudiana* of Duchassaing and Michelotti. I have been unable to find anything in the description or figures furnished by these authors (*Supplément au Mémoire sur les Coralliaires des Antilles*, p. 162, pl. IX, figs. 1, 2—*Memorie della Reale Accad. Scienze di Torino*, Ser. Sec., XXIII, 1866) to distinguish their species from *Oculina speciosa*, nor does it appear to me to be distinct. The characters upon which the form is separated are exceedingly trivial, and well within the amount of variability which is presented by individual specimens of nearly all the species of Oculina. I further believe that *O. coronalis*, and possibly also *O. recta*, will have to be united with *O. speciosa*.

Isophyllia australis ? Edwards and Haime.

Three specimens from the North Rock, doubtfully indentified with this species.

Isophyllia fragilis ? Dana.

I am unable to satisfy myself as to the positive existence of this species in Bermuda, although Quelch refers to a single specimen having been obtained there by the Challenger party. This authority doubtfully refers one of the forms figured by Pourtalès (*op. cit.*, pl. VII, fig. 3) as *I. dipsacea* to Dana's species, but from an examination of a number of Bermudian specimens which agree absolutely with Pourtalès's figure I am fairly convinced that this identification is incorrect. The specimens do certainly not agree sufficiently with Dana's description, and if they are not the types of a distinct species, then they represent probably only a certain phase of development of *I. dipsacea*, as is indicated by Pourtalès.

Isophyllia dipsacea, Dana.

Three specimens, from Castle Harbor.

Isophyllia strigosa, Duchassaing and Michelotti

A number of specimens, from Harrington Sound, which agree with the description of this species. I am doubtful as to the species being distinct from *Isophyllia dipsacea*; possibly, however, some of the varieties (so-called) of the latter species figured by Pourtalès are really members of this species. Its principal distinguishing characters appear to be the thinner and more irregular septa, and the terminal cleft that indents or separates the septa of opposing calyces where they cross the common wall. It also presents a more bristling appearance than *I. dipsacea*.

Isophyllia Guadeloupensis, Pourtalès.

One specimen. This appears to be a good species, although by Quelch it is referred to *Isophyllia strigosa*.

In addition to these forms Quelch enumerates *Isophyllia* (*Symphyllia*) *marginata, I. cylindrica,* and *I. Knoxi,* all of Duchassaing and Michelotti, as having been obtained at the

Bermudas, but I have failed to detect any specimens among our collections which can be confidently referred to these species. On the other hand, I find one or two forms which I have not yet been able to identify with any described forms.

Siderastræa galaxea, Ellis and Solander.

Abundant on the shoals of Gallows Island, near the mouth of Flatts Inlet, where the colonies come to within about two inches of the surface; also on the borders of Harrington Sound.

Porites clavaria, Lamk.

Two specimens, dredged in Harrington Sound.

Porites astræoides, Lamk.

We found this species very abundantly along the outer reef, especially on the flats of the North Rock, where it is the dominant form of coral. The species appears to have been overlooked by the Challenger party, and indeed, the only reference that I have been able to find indicating the occurrence of this common West Indian form among the Bermudas is contained in Mr. Rathbun's list of the species of Porites in the United States National Museum (Proc. U. S. National Museum, 1887, p. 354).

Mæandrina labyrinthica, Ellis and Solander.

Three specimens, from the North Rock.

Mæandrina strigosa, Dana.

This form is represented by large, sub-globose specimens, one of which, obtained through purchase, and probably from Castle Harbor, has an exceedingly attenuated base of attachment. The corallum is thus openly turbinate, or even pediculate, and exhibits in its regular scalariform outline the successive stages of outward development.

Diploria cerebriformis, Lamk.

This species is exceedingly abundant in the shoals lying to the leeward of the marginal reef, where its huge hemispherical or reniform masses of bright orange, measuring as much as four or five feet in diameter, can be distinctly seen through the transparent waters at depths of from six to fifteen or twenty feet. I cannot say how much further down the species extends. It is equally abundant in Castle Harbor, where it is largely instrumental in building out the shore-platform which, at a moderate distance from the shore, descends vertically into deeper water. When attached by a contracted base, the brain-coral may be readily removed from its moorings; but where the base is largely coextensive with the under-surface of the corallum the difficulties of removal are very great, necessitating much undercutting with a chisel. The largest specimen obtained by us measured about 28 inches across; our efforts to dislodge a specimen about four feet in diameter proved unsuccessful.

Diploria Stokesi, Edwards and Haime.

We obtained a number of specimens of this species in Castle Harbor, and through presentation; for the latter my thanks are due to Miss A. Peniston, of Peniston's. The habitat of the species, as far as I am aware, has not hitherto been noted. Edwards and Haime in their description of the species (*Hist. Nat. des Coralliaires*, II, p. 403, pl. D, fig. 3) state " *Patrie inconnue*." I believe it may be assumed that this species is the form described and figured by Knorr as *Madrepora labyrinthiformis* (*Deliciæ Naturæ Selectæ*, I, p. 18, Pl. A 4, fig. 1). In our collections we have a closely related, and possibly identical, species, which assumes a ring form, and in which the peculiar calycular hollows of *D. Stokesi* run out into parallel transverse grooves on the inner side of the ring.

ALCYONARIA.

The gorgonians are abundant in the waters inside of the bounding reef, whence nearly all our specimens were obtained. A few were nipped up on the south side of Castle Harbor, and at the passage way conducting from the north into that body of water.

Rhipidogorgia flabellum, Valenciennes,

The purple variety of this species is abundant more particularly in the northern waters, both near the outer reef and on the shallows known as Devonshire Flats. We failed to obtain any of the yellow forms, and I am not positive that this variety has ever been found at the Bermudas.

Gorgonia (Plexaura) purpurea, Pallas.

Gorgonia (Plexaura) flexuosa, Lamouroux.

This species, of which we obtained several specimens, is, I believe, without doubt the *Gorgonia anguiculus* of Dana (U. S. Exploring Expedition, Zoophytes, p. 668). It is referred to under Lamouroux's name as a member of the Bermudian fauna in Dana's "Corals and Coral Islands," p. 114, 1872.

Gorgonia (Plexaura) homomalla, Esper.

Gorgonia (Plexaura) multicauda, Lam.

Gorgonia crassa, Ellis and Solander.

G. vermiculata, Edwards and Haime.

The exact limitations and synonymy of this species are difficult to make out, but as far as my studies have permitted me to analyze the forms above indicated from the rather insufficient or deficient descriptions that have been furnished by their authors, they appear to represent an identical form. As such I have accordingly referred them in this list.

Gorgonia (Plexaura) dichotoma, Esper.

A single specimen, measuring about a foot and three-quarters in height, with the main stems somewhat over a half-inch in diameter.

Gorgonia (Eunicea) pseudo-antipathes, Lam.

One much branched specimen, and another, slightly differing, which appears to belong to the same species.

Pterogorgia acerosa, (?) Pallas.

A single specimen of a large Pterogorgia, entirely stripped of coenenchyma, and measuring about two and a-half feet in height, was obtained through purchase at the Crawl. The axial skeleton is yellowish, or of the color of earth. The terete branches are much more broadly spreading than in *P. setosa*, and unite into a common basal stalk which is upwards of two inches in thickness. The pinnules are very numerous, exceedingly slender, and pendulous, giving to the whole organism the decided appearance of a weeping-willow.

I have not been able to satisfy myself as to the exact affinities of this species. It appears to differ broadly from the common purple sea-feather of the West Indies, and does not have the depressed branches which are assumed for Esper's *Pterogorgia accrosa*. It is, however, with little doubt one of the forms that are included by Pallas in his *Gorgonia accrosa* (*Quercus marina Theophrasti*), and may be the one that is referred to by Milne-Edwards as *Pterogorgia Sloanei*.

Of the species of gorgonians above enumerated Dana indicates *Rhipidogorgia flabellum*, *Gorgonia flexuosa*, *G. homomalla*, and *G. crassa* as coming from the Bermudas ("Corals and Coral Islands," p. 114). I find no mention in any more recent work of the occurrence there of either *Gorgonia pseudo-antipathes* or *G. dichotoma*. On the other hand, we failed to obtain the *Pterogorgia Americana* mentioned by Dana.

ACTINIARIA.

For the following contribution to the "Actinology of the Bermudas" I am indebted to Prof. J. Playfair McMurrich, who has carefully examined and studied all the specimens contained in our collection. These were not very numerous, but still sufficiently so to present a number of interesting points in special morphology and geographical distribution. The observations here recorded have appeared in advance in the Proceedings of the Academy of Natural Sciences of Philadelphia.

THE ACTINOLOGY OF THE BERMUDAS.

BY

PROF. J. PLAYFAIR MCMURRICH.

I recently received from Professor Heilprin a number of actinians which he had collected in the summer of 1888, during a visit to the Bermuda Islands. They were entrusted to me for identification and study, and I gladly availed myself of the opportunity thus afforded of comparing the actinian fauna of the Bermudas with that of the Bahamas, which I had previously studied.* I may state here that, so far as can be judged from the material studied, there is very great similarity between the two faunas, most of the species from the Bermudas occurring also either in the Bahamas or in the West Indian Islands. Unfortunately, it was impossible to adopt the best methods of preserving the material obtained in the Bermudas, the expedition to the islands having been undertaken mainly for geological purposes, and consequently the specific relationships of some of the forms could not be determined with perfect certainty.

SAGARTIDÆ.

Aiptasia. sp? (Pl. 10, figs. 1 and 2.)

In the collection were four specimens of a form which I refer to the genus *Aiptasia*, inasmuch as in the majority of respects they resemble forms of that genus, although it was impossible to ascertain the presence of an equatorial row of cinclides owing to the ectoderm having been almost completely macerated away. Nematocysts were quite abundant in the macerated substance contained in the inter- and intra-mesenterial, chambers, but it was not possible to be certain that they belonged to acontia, though such was probably the case.

The specimens were about 1 cm. in length and 0.65 cm. in diameter. The color, as ascertained from the alcoholic material, is in the upper one-third of the column and in the tentacles grass-green, while the rest of the column presents the dirty grayish-brown color frequent in alcoholic specimens. About one-third of the way down the column each specimen presents a well marked constriction, below which the column is cylindrical, while above it gradually expands, the disc not being at all infolded in contraction. The base is evidently adherent, but in two of the specimens it is much smaller than

*See Journal of Morphology, vol. iii. This paper is now in print and will shortly appear.

the column, and is almost covered by the infolding of the column walls over it; this apparently, however, is an abnormal condition. The column is smooth, and no traces of cinclides could be seen as stated above. Sections (Pl. 10, fig. 2) show that the mesogloea is thin throughout, and that the circular muscles (cm) are only feebly developed. There is a special sphincter (sp) imbedded in the mesogloea, immediately below the margin, and, though not very powerful, is yet quite apparent. The only species of *Aiptasia* in which such a sphincter has been observed as yet is *A. pallida* of our Eastern coast. Immediately below this the mesogloeal muscular processes which support the circular muscles are weak, but further down they enlarge gradually and form a second sphincter (sp') similar to what has been described by R. Hertwig* in *Leiotealia nymphoea*. It is to the presence of this lower sphincter that the contraction of the column mentioned above is due.

The tentacles are 48 in number and are arranged in four cycles. They are strongly entacmaeous, and are not infolded during contraction. Those of the first cycle measure 1.1 cm., and those of the outermost cycle 0.3 cm. The ectodermal and endodermal muscular processes are present, but do not call for a special description. The disc is flat and the stomatodaeum is without well-marked gonidial angles; sections show that the grooves are hardly developed.

The mesenteries are in four cycles. The six pairs of the first cycle are alone perfect; those of the second cycle are shorter but provided with well-developed longitudinal muscles, while neither those of the third nor those of the fourth cycle have the longitudinal muscles, the members of the latter cycle not projecting above the surface of the endoderm. The parieto-basilar muscles seem to be wanting, or at least have no marked mesogloeal processes. The reproductive organs are borne by the mesenteries of the second cycle, and also by those of the first cycle (except by the directives) below the internal opening of

---

*R. Hertwig—Report on the Actiniaria. Zoology of the Voyage of H. M. S. Challenger. Vol. vi. Pt xv. 1882.

the stomatodæum. This is the only Sagartid, with the exception of *A. pallida,* in which I have observed reproductive organs
on the mesenteries of the first cycle, and it is a case of considerable importance inasmuch as it necessitates an alteration in
the definition of the family Sagartidæ as given by R. Hertwig.[*]
One of the essentials of the family is that "the principal septa,
or septa of the first order, only are perfect and at the same
time sterile." The last portion of this statement, though true
for the majority of Sagartidæ, fails in the case of the Aiptasiæ
mentioned. It is not possible to separate *Aiptasia* from the
Sagartidæ; the members of the genus possess acontia, cinclides,
the primary mesenteries alone perfect, and a mesodermal muscle in some cases, and these must be considered as the chief
characteristics of the family.

As regards the species to which the form under consideration
belongs, the probabilities are that it is identical with *A. pallida*
of our Eastern coast, since in its anatomical peculiarities it
agrees very closely with that form. The impossibility however
of ascertaining the coloration, and what is of much more importance, the occurrence and arrangement of the cinclides,
has prevented a certain identification, and I have preferred to
leave the species in doubt.

ANTHEADÆ.

Condylactis passiflora. Duch. and Mich. (Plate 10, fig. 3.)

Several specimens were obtained of a large form, measuring
3.3-2.3 cm. in height and 2.6-3.8 cm. in diameter when preserved, which resembled in coloration, external characters, and
for the most part in internal structure also, the West Indian
form *Condylactis passiflora.* In the alcoholic specimens the column is of a brick-red color wherever the ectoderm has been
preserved, and the tentacles are grass-green, this color evidently being due to the enormous number of zooxanthellæ
contained in the endoderm. Professor Heilprin informs me to
the best of his recollection the tentacles in the living specimens

---

[*] *Loc. cit.*

were as a rule tipped with crimson.   In a separate bottle is a
single specimen evidently identical with the others, and accompanying it is a note stating that the column was red and the
tentacles brown.   This specimen was found freely floating near
the surface, but had evidently become detached, as its base
shows that normally it is an attached form.

The ectoderm having been macerated away, the outer surface
of the mesogloea is exposed to view, and is seen to be divided
by fine longitudinal and transverse grooves into small quadrangular areas.   These grooves are continued over the limbus
upon the surface of the base, the longitudinal grooves there
becoming radiating and the transverse ones concentric.

The only character which is markedly different from what
occurs in the West Indian specimens of the species is presented
by the longitudinal muscles of the mesenteries.   The middle
portion of a section through the muscle-band presents an
appearance quite similar to that seen in the West Indian
form, and the internal edge is also the same, the long mesogloeal processes terminating abruptly, and being followed by
smaller processes which extend to the commencement of the
reproductive region of the mesentery; but toward the insertion of the mesenteries into the column wall the arrangement
is slightly different (Pl. 10, fig. 3).   In the Bahama specimens
the mesogloea between the outer edge of the muscle-band and
the insertion of the mesentery into the column wall is thin,
and the muscle-band gradually thins out externally.   In the
Bermuda forms, however, the longitudinal muscle begins abruptly, and the mesogloea external to it is thick, with short,
stout muscle processes; or, as in the directives, with the
muscle cells, instead of appearing to cover processes, presenting rather the appearance of here and there dipping down
slightly into the mesogloea.

It is not probable, however, that this slight difference is to
be regarded as specific, and since in other respects there is
almost exact correspondence, the Bermuda forms must be considered identical with those from the Bahamas.

PHYLLACTIDÆ.

Some points of considerable importance as regards the characteristic structure of the members of this family have been obtained from the study of the two forms which I include here within it. The family was established by Andres* for forms in which the disc is furnished towards the center with simple tentacles and towards the periphery with foliaceous fronds. In one of the forms about to be described the fronds are replaced by short digitiform tentacles arranged in a single cycle, but nevertheless it agrees in other structural points with *Oulactis*, and I have therefore found it necessary to alter the definition of the family, placing importance on internal anatomical structures rather than upon external characteristics.

In the first place, in the *Oulactis* about to be described, and in *Diplactis*, as I propose to name the genus to which the form with tentacles replacing the fronds will be referred, a sphincter of the diffuse type is present, but instead of being situated upon the column wall below the margin, it occurs internal to the margin, between the inner tentacles and the peripheral fronds or tentacles. In *O. flosculifera* from the Bahamas this sphincter was not observed, but was probably overlooked in the single specimen I obtained for study, and none of the preparations which I still possess include the region in which the sphincter should occur. Secondly, in the two species of *Oulactis* which I have studied, and in the *Diplactis*, the gonidial grooves are very deep, and are prolonged a considerable distance below the inner margin of the stomatodæum; the histological structure also of the ectoderm lining the grooves differs slightly from that of the general surface of the stomatodæum, it is not thrown into folds as it is elsewhere, and the mesoglœa of the grooves is thickened.

I would define the family Phyllactidæ as follows:—Actininæ in which the disc is furnished with simple tentacles towards the center and with a cycle of short digitiform tentacles or

*A. Andres. Le Attinie. *Fauna und Flora des Golfes von Neapel, Monographie* ix. 1883.

more or less foliaceous fronds towards the periphery; a sphincter of the diffuse type occurs upon the inner surface of the disc between the inner tentacles and the outer tentacles or fronds; and the stomatodæum is provided with two deep gonidial grooves, which are prolonged some distance below the inner extremity of the stomatodæum.

The family Phyllactidæ was placed by Andres in the sub-order (family) Stichodactylinæ, the fronds being considered homologous with tentacles. I have here ventured to remove the family to the sub-order Actininæ, and it will be necessary to furnish my reasons for such a change. The tentacles must necessarily be considered outgrowths of the disc, since structurally they resemble it closely while differing greatly from the column. Are the fronds also disc structures?

The question turns upon what we shall consider to be the limit between the disc and the column. The majority of authors have taken a more or less distinct fold of the body wall, the margin, frequently furnished with conspicuous acrorhagi, to be the boundary, and certainly in many cases there seems to be a marked difference on either side of this fold. Thus, the column may, as in *Bunodes* and *Phymactis*, be turberculated as far as the margin, but beyond this the tubercles cease, and there is apparently a decided difference between the region below and that above the limiting fold.

In the Sagartidæ and Paractidæ there is imbedded in the column wall below the margin a sphincter muscle. In other forms, however, such as the Bunodidæ, which possess a circumscribed endodermal sphincter, that structure lies internal to the margin. If we assume with the Hertwigs that the sphincter is a columnar structure, its situation in the Bunodidæ would indicate that the margin is not the boundary between the disc and column.

Neither the margin nor the sphincter, however, can be considered the morphological boundary of the disc, since both seem to vary somewhat in position. The true criterion is to be found in the difference of histological structure presented by the disc

and column ectoderm. This layer in the disc possesses ectodermal muscle-cells and a nerve-layer, which structures are
absent in the column. The tentacles resembling the disc in
structure are to be considered outgrowths of that region, and
passing outward from these one finds that the characteristic
structures of the disc gradually fade out and are lost. It is
impossible to say just where the change is completed, but the
region in which it occurs must be considered the boundary between the disc and column. In *Bunodes tæniatus* and *Aulactinia
stelloides* I find that the sphincter muscles lie beneath the outer
border of this indifferent region, and are consequently to be regarded as columnar structures.

In the Phyllactidæ the sphincter muscle lies between the
tentacles and the fronds, and although the ectoderm in the region in which it occurred, and in the area between the fronds
or their representatives and the margin was completely macerated away in the forms studied, yet reasoning from the relations of the sphincter in other forms we must conclude that the
region between the margin and the base of the tentacles is
columnar, and that the fronds and outer digitiform tentacles
are column structures, perhaps comparable to acrorhagi, and
cannot be considered homologous with tentacles. Accordingly,
only one tentacle belongs to each intra-mesenterial space, and
the Phyllactidæ must be referred to the sub-order Actiniæ.

Andres, in the introduction to his Monograph, notes the fact
that the margin does not always mark the boundary between
the disc and the column. He proposes the term "collar" to
denote the portion of the column internal to the margin.
Gosse's term "fosse" is not applicable in all cases, as for instance in *Condylactis*, where the region does not form a depression, but is horizontal.

Oulactis fasciculata. n. sp. (Pl. 10, fig. 5.)

By this name I denote three specimens in various degrees of
contraction, the largest of which measured about 1 cm. in
height and 1-2 cm. in breadth. The color, as ascertained from
alcoholic specimens, is in the lower part of the column a gray-

ish-brown, similar to what is frequently seen in preserved Actininæ, while the upper part of the column and the fronds are of a grass-green, the tentacles resembling somewhat the lower part of the column, but having a distinctly greenish tinge.

The column is provided in its upper part with about 48 vertical series of tubercles, probably verrucæ, there being about five or six in each series, and is thrown into numerous transverse folds, the result of contraction. The mesogloea, when exposed, appears to be raised into numerous minute elevations, whereby the surface acquires a finely punctured appearance.

The tentacles are moderately long, simple and pointed at the extremity. They appear to be arranged in two cycles, and from a necessarily uncertain count I estimate their number to be about forty-eight. Their ectodermal longitudinal muscle layer is well developed, being arranged on long slender mesoglœal processes. The fronds (Pl. 10, fig. 5, fr.) are small, yet occupy the entire width of the area between the tentacles and the apparent margin. They consist of hollow evaginations of the disc, arranged in bunches. I could not determine with certainty their number in any of the specimens, but there are probably twenty-four of them in all. A well-defined margin is present.

Immediately external to the bases of the tentacles, and lying between them and the fronds there is an endodermal sphincter (sp) fairly well developed. Immediately external to it, in the region occupied by the fronds, and for a slight distance down the column-wall below the margin, there are no muscle processes, but further down they do occur, forming what might be termed a second sphincter, though it is by no means well developed. The surface of the disc between the tentacles and the mouth is deeply depressed, so that a fosse is formed around the peristome. The mouth is large. Sections show that over the general surface of the stomatodæum the mesoglœa is very thin, and upon the ectodermal surface gives rise to numerous more or less regularly arranged fine processes, over which the ectoderm passes so as to be thrown into numerous folds. The gonidial grooves are deep, and are prolonged some distance below the rest of the stomatodæum. Its mesoglœa is much thickened, and is devoid of processes upon its ectodermal surface, being thus strongly contrasted with that of the stomatodæum. In its histology the ectoderm of the groove also differs from that of the general stomatodæum, the glandular cells being evidently fewer in number, but the preservation of the specimens was not sufficiently perfect to permit the details to be made out.

There are altogether twenty-four pairs of mesenteries, twelve of which are perfect. The six primary pairs are united with the stomatodæum to a greater extent than are the six secondaries, and the two pairs of directives have a much more extensive union than any of the other primary mesenteries, owing to the great prolongation of the gonidial grooves. The longitudinal muscle processes form a strongly projecting though rather narrow band, the edges of which are sharply defined, the processes being of equal length throughout the muscular area, and diminishing abruptly towards the sides. The mesoglœa of the portion of the mesenteries external to the muscle bands is rather thick, and there is a strong parieto-basilar muscle. Apparently only the mesenteries of the third cycle, *i.e.* the imperfect mesenteries, are gonophoric, but my preparations do not allow of certainty on this point.

A few remarks are called for concerning the relationship of this species. I was at first tempted to identify it with *O. formosa*,[*] but further consideration led me to separate it as a new species. The fronds differ markedly from those of other species of *Oulactis*. In these they have been described as being "*chicoracés*," a term which cannot be applied to the fronds of *O. fasciculata*. In it they consist of bunches of finger or club-shaped hollow processes, the various processes of each bunch being united by their base but distinct above. This arrangement suggested the specific term which I have em-

---

[*]Duchassaing and Michelotti—*Mem. Reale Accademia di Torino.* 2nd Ser. xix, 1860, and xxiii, 1866.

ployed, and I think is of sufficient importance to warrant the formation of a new species. It was a question whether a new genus should be instituted, as Verrill[1] has done in the case of *Lophactis ornata*, but there is such close agreement with the Bahaman *O. flosculifera* as regards the internal structure, the number of perfect mesenteries, and the distribution upon the mesenteries of the reproductive organs, that such a proceeding was considered unnecessary. It seems probable that the genus *Lophactis* should be fused with *Oulactis*.

DIPLACTIS. Gen. nov.

I propose this generic name for two species, one of which is described below, which do not seem to be referable to any of the genera now recognized. The genus may be briefly defined as follows:—Phyllactidæ in which the fronds are represented by a single cycle of short digitiform tentacles and in which all the mesenteries except those of the first cycle are gonophoric. The term *Diplactis* has been chosen as indicating the tentacular appearance of the fronds, from which it seems as if there were two series of tentacles, an inner and an outer (*diplous*, double, and *aktis*, a ray).

In the Supplementary Report on the Actiniaria collected by the "Challenger," R. Hertwig decribes a form whose locality is unknown, which he refers to Gosse's genus *Hormathia*. It is very similar to the form about to be described from the Bermudas, and there can be no doubt that though specifically distinct the two must be referred to the same genus. Gosse's *Hormathia*[2] was described from a single specimen brought up on a deep-sea fishing line, and attached to the shell of a living *Fusus*. It was characterized by possessing slightly below the margin about ten spherical protrusions. I do not think it is possible to associate in the same genus with this either the Bermuda *Diplactis* or Hertwig's *Hormathia*. In the first place in both these forms the bodies near the margin are digitiform and not spherical; and secondly, these bodies are situated not

---

[1] Verrill,—Trans. Conn. Acad. Vol. i, 1868.

[2] P. H. Gosse. "Actinologia Britannica." London. 1860.

below the margin, but internal to it. In all the Bermuda specimens, of which there are quite a large number, a well-marked
margin is present and Hertwig describes in his form a fold of the
column-wall which must be considered equivalent to the margin of the other species. Although the margin cannot be considered of importance as marking the boundary of the disc,
yet it is a structure of frequent occurrence and must be taken
account of. Structures that occur on the column-wall below
it, as in *Hormathia*, cannot be considered identical for systematic purposes with others which invariably lie above or
internal to it, and are not quite similar in form.

It must be noticed that Haddon has recently referred to
Gosse's *Hormathia* a form* altogether different from that assigned to it by Hertwig. The correctness of Haddon's identification is quite as doubtful as Hertwig's, if not a little more
so. The form is certainly a Sagartid, and probably a *Phellia*,
it being stated that it is very similar to Hertwig's *Phellia
pectinata*.

Hertwig refers his *Diplactis* (*Hormathia*) *delicatula* to the family *Antheadæ*, on account of the diffuse endodermal nature of
sphincter. The situation of the muscle and other characters
make it evident that *Diplactis* should be associated in the same
family with *Oulactis*. In the Phyllactidæ as here limited we
have several grades of complication of the fronds. In *Diplactis*
their structure is exceedingly simple, being simply digitiform
in *D. Bermudensis* and club-shaped in *D. delicatula*. In *O. fasciculata* they are somewhat more complicated, and from this the
passage is easy to Verrill's *O.* (*Lophactis*) *ornata*, and from this
to the very complicated structure seen in *O. flosculifera*.

Diplactis Bermudensis.   n. sp.   (Pl. 10, figs. 4 and 6, Pl. 11, figs. 1 and 2.)

A number of specimens of the form for which I propose this
name were obtained. The majority were in a partially contracted condition, but apparently the power of contraction is
not fully developed, as in none were the tentacles completely

---

* A. C. Haddon.—On two species of Actiniæ from the Mergui Archipelago.—
Journ. Linn. Soc. Vol. XXI. 1888.

concealed. The average height of the specimens is about 1·5 cm. and the breadth nearly the same, and thus *D. Bermudensis* is decidedly smaller than *D. delicatula*. The ectoderm has been almost entirely macerated away, so that the external surface of the mesoglœa is exposed to view. This presents numerous transverse folds due to contraction, but in addition fine linear depressed striæ are present, both horizontal and longitudinal, dividing the surface into numerous rows of small quadrangular elevations visible to the unaided eye. The color throughout is a dirty-green.

The base is adherent, flat, and about the same size as the column. It is marked by radiating and concentric striæ, continuations of the longitudinal and horizontal striæ respectively of the column. No verrucæ or tubercles occur on the column, though the quadrangular areas produced by the striæ are slightly more prominent toward the margin. This is well marked and smooth, and is separated from the tentacles by a deep fosse (collar), near the bottom of which are about 12 short digitiform fronds about 1 mm. in height. (Pl. 10, fig. 4p, and Pl. 11, fig. 1.)

Between these fronds and the tentacles there is upon the endodermal surface of the collar a sphincter (Pl. 10, fig. 4, sp) of the diffuse type, which differs markedly from that of *D. delicatula*, the mesoglœal processes being much more delicate, and anastomosing somewhat in their proximal portions (Pl. 11, fig. 2). The circular muscles of the column wall external to the digitiform fronds are fairly prominent, and are continued the entire way down the column, not enlarging however to form a second sphincter.

The tentacles are simple, conical, and of moderate length, and are arranged in four cycles, their number being probably 96. By actual count they seemed to vary somewhat, usually falling below that number, but the discrepancies are probably due to the difficulty of making a correct enumeration. They possess well-developed mesoglœal processes for the support of the ectodermal muscles. The disc is deeply folded in, in-

ternal to the tentacles, so as to form a deep fosse around the
peristome (Pl. 11, fig. 1), which, however, does not rise above
the level of the margin.   I was not able to distinguish in any
of my preparations the delicate mesoglœal processes of the disc
which support the ectodermal muscle cells in *D. delicatula*, but
it is possible that they had been macerated away.

The mesoglœa of the stomatodæum is raised upon its ectodermal surface into prominent but rather delicate ridges, over
which the ectoderm is folded.   The gonidial grooves are deep,
and as in *Oulactis* are prolonged below the level of the internal
opening of the stomatodæum, and have the mesoglœa thickened.   In *D. delicatula* Hertwig describes the gonidial grooves
as being hardly marked in the stomatodæum, and if this is
found to be an invariable characteristic, it will be necessary to
alter slightly the definition of the Phyllactidæ given above.
The depth of the grooves, and their prolongation downwards,
are so marked in the other members of the group that I have
examined, as to suggest that the apparent shallowness in the
specimen examined by Hertwig may be due to distortion.

The primary and secondary mesenteries are perfect throughout the whole length of the stomatodæum; the tertiaries are
perfect in their upper part, but lower down separate from the
stomatodæum; while the fourth cycle consists entirely of imperfect mesenteries.   The directives are attached throughout
a greater part of their length than are any of the other mesenteries, owing to the prolongation of the gonidial grooves.   The
longitudinal muscles in the upper portion of the mesenteries
form a low band, covering the greater portion of the non-gonophoric region of the mesentery; internally the muscle processes
end rather abruptly, but externally they gradually diminish
in size.   In the lower part of the mesentery, below the level of
the stomatodæum, the arrangement of the muscle processes is
very different (Pl. 10, fig. 6).   Throughout the greater portion
of the non-gonophoric region of the mesentery they are very
small, but as the gonophoric region is approached they suddenly increase in size, forming a strong projection, and then

just as suddenly diminish again, the projection being of slight extent. The parieto-basilar muscles are well developed and form conspicuous folds. Both external and internal mesenterial stomata are present (Pl. II, fig. 1). All the mesenteries with the exception of those of the first cycle are gonophoric.

The differences between *D. Bermudensis* and *D. delicatula* may be briefly enumerated as follows:

D. Bermudensis.	*D. Delicatula.*
Tentacles 96.	Tentacles 160.
Fronds digitiform, about 12 in number.	Fronds dilated at the extremity, about 42 in number.
Mesoglœal processes of sphincter muscle rather delicate, anastomosing slightly.	Mesoglœal processes of sphincter stout, not anastomosing.
Ectodermal muscle processes of disc wanting (?)	Ectodermal muscle processes of disc long and delicate.

PHYMANTHIDÆ.

Phymanthus crucifer. (Les.) Andres.

A single specimen of this species was obtained. I have nothing to add to the statements already made regarding it in my paper on the Bahama Actiniaria.

ZOANTHIDÆ.

Zoanthus flos-marinus. Duch. and Mich. (Pl. II, figs. 3 and 4.)

A large number of specimens of this species were obtained, and inclosed with them was a label stating that they were collected at Shelly Bay and Tucker's Town. In general appearance they resemble *Z. sociatus* from the Bahamas, the individuals as in that species forming stolon-like prolongations from which new individuals bud; their structure, however, shows them to belong to a different species. The colonies are, according to the accompanying label, 4 to 5 cm. in breadth. The individual polyps in the preserved condition measure 1-2 cm. in height, and in breadth at the upper end 0·5 cm. the lower portion and stolons measuring about 0·25 cm. All are strongly contracted, a small depression being the only indication of where the entrance into the interior is situated. The color, according to the inclosed label, was "spinach-green," but this must be taken as applying only to the upper part of the column, the lower part and the stolons being brown or sand-colored. The disc was "apple-green," and the tentacles green.

The column upon the outside is covered by a cuticle, in which are sparingly imbedded foreign bodies. The ectoderm is separated from the cuticle by a layer of mesoglœa, and consists of cells arranged in groups separated by partitions of mesoglœa, but not showing the degeneration which occurs in *Z. sociatus*. The mesoglœa is comparatively thick, and consists of a homogeneous matrix containing (1) numerous anastomosing spaces more or less filled with cells, and (2) granular cells which give rise to delicate processes which enter into connection with other granular cells, and with the spaces just mentioned, and with the ectoderm and the endoderm. Some suggestions regarding the origin and function of these structures will be found in connection with the description of *M. tuberculata* which follows. The endoderm of the column is low, and consists of more or less spherical cells, usually containing zooxanthellæ. A delicate layer of muscle fibres arranged circularly occurs between the endoderm and the mesoglœa.

At the upper part of the column a well-developed double sphincter muscle occurs, imbedded in the mesoglœa. It is stronger than that found in *Z. sociatus*, and more nearly resembles that described by Erdmann[1] and Hertwig[2] in *Z. Danæ*.(?)

The tentacles, according to the brief notes taken of the living specimens, are "short, 50-60 in number, in three rows." My preparations, however, show that the last statement is erroneous, the tentacles being arranged in two cycles only. Their ectoderm is not imbedded in the mesoglœa, nor is there a cuticle covering it. The ectodermal muscle processes of the

[1] A. Erdmann. *Ueber einige neue Zoantheen.* Jen. Zeit. XIX. 1885.

[2] R. Hertwig. Supplement to report on the Actiniaria. Zoology of the Voyage of H. M. S. Challenger. Vol. XXVI. 1888.

mesoglœa are fairly developed, and immediately below them are to be seen, imbedded in the mesoglœa, peculiar granular pale yellowish-green cells, the protoplasm of which, with the exception of the nucleus, does not stain with carmine. Otherwise the mesoglœa is homogeneous. The endoderm is thick, and is richly supplied with zooxanthellæ. In structure the disc resembles the tentacles, possessing, like them, the peculiar yellowish-green granular cells.

The mesoglœa of the stomatodæum is homogeneous. I cannot make any statements as to the histology of the ectoderm of this region, as it had macerated into a mass of a characteristic appearance which cannot easily be described. Transverse sections show that the gonidial groove, to which the macro or ventral directives are attached, is very shallow, and indeed can hardly be said to exist.

The mesenteries are arranged on the microtypus.* Their mesoglœa is for the most part very thin but thickens towards the base where it contains a canal. (Pl. 11, fig. 4, bc.) A second canal, circular in section and packed with cells, occurs in the thin region, the mesoglœa splitting to form its walls. The muscle layers are only slightly developed.

A very peculiar arrangement occurs in connection with the mesenterial filaments of the perfect mesenteries. Immediately below the stomatodæum the mesenterial filament is triradiate (Pl. 11, fig. 3), the central ray being short and stout, the lateral rays longer and recurved. The epithelium covering the central ray and that face of the lateral rays which looks towards it resembles in structure that of the stomatodæum. The outer surface of the lateral rays is, however, covered with cells similar to those which line the general surface of the mesentery. In a section which passes through the stomatodæum a little above its extremity, the intervals between the perfect mesenteries is occupied by macerated tissue resembling the ectoderm of the stomatodæum. Apparently, it lines the surfaces of the mesen-

* See Erdmann, *loc. cit.*

teries for a short distance outwards from their point of attachment to the stomatodæum, and also the outer surface of the latter for a short distance above its inner opening. It looks as if
the ectoderm of the stomatodæum were reflected upwards, so as
to cover its endodermal surface and the adjacent surfaces of
the perfect mesenteries. Further down (Pl. 11, fig. 4) the two
lateral processes of the mesenterial filaments disappear, the central one alone persisting. It is evidently the " glandular streak "
of the filament. The cells which cover the surface of the mesentery for some distance outward from this towards the column-
wall are very peculiar. (Pl. 11, fig. 4di.) They form a layer
much thicker than that formed by the ordinary endodermal
cells, and are loaded with green granules, closely packed
together so that to the naked eye the region occupied by this
layer is of that color. Foreign bodies of organic nature are
imbedded in the cells, sometimes being surrounded by a number of cells containing no granules, or occasionally imbedded
in the mesoglœa.

In unstained specimens, when the animal is laid open by a
longitudinal incision, this region of the mesenteries is very
distinct on account of its rich green color. When the loose
cells of the green area are scraped away with a scalpel and
examined, they are seen to be of a very irregular shape (sug-
gesting a power of amœboid movement), and to contain numerous green globules, much smaller than the zooxanthellæ,
darker in color, and homogeneous in structure. Amongst the
cells are numerous zooxanthellæ, and there are also numerous
spherical refractive bodies, apparently of a fatty nature and
with a slightly-greenish tinge, as well as the foreign bodies
already mentioned as seen in the section, and very numerous
delicate acicular silicious spicules.

The occurrence of these spicules and organic foreign bodies
in the cells of this region is very strong evidence in favor of
the supposition that they have a digestive function. The
green globules may be the products of digestion. If this be
the case it is exceedingly interesting, as indicating a method of

digestion in the Zoantheæ somewhat different from what is usually described as occurring in the rest of the Actiniaria.

None of the specimens examined possessed sexual organs. There were about 24–26 pairs of mesenteries in the specimens examined.

I have identified this form with *Z. flos-marinus* of Duchassaing and Michelotti, with the imperfect description of which it agrees fairly well. In many respects it comes near *Z. sociatus*, but differs markedly from it in others; such for instance, as in the nature of the ectoderm and in the form of the sphincter-muscle, so that it must be regarded as distinct. From the only *Zoanthus* hitherto described from the Bermudas, *Z. Danæ* (?) of Hertwig[1] it is readily distinguished by the absence of any distinct line of demarcation between the upper and lower portions of the column.

Mammillifera tuberculata (Gray) (Pl. 11, figs. 5 and 6.)

> *Isaurus tuberculatus*—J. E. Gray. 1828.
> *Zoanthus tuberculatus*—Duchassaing and Michelotti. 1860.
> *Antinedia tuberculata*—Duchassaing and Michelotti. 1866.
> *Zoanthus (Monanthus) tuberculatus*—Andres. 1883.
> *Antinedia Duchassaingi*—Andres. 1883.

This form was first described by J. E. Gray,[2] from specimens in the British Museum whose locality was unknown. He adopted for the genus Savigny's name *Isaurus*. In 1860, Duchassaing and Michelotti rediscovered it, and, though apparently unacquainted with the earlier description of Gray, applied to it the same specific name, but placed it in the genus *Zoanthus*, on account of the absence of sandy incrustations on the column walls. In their second paper these authors, placing importance on the tuberculation of the column walls, erected for its reception the genus *Antinedia*. Andres, in his most useful monograph, has assumed that the form described by Gray

---

[1] R. Hertwig. Supplement to report on the Actiniaria of the Challenger Expedition. 1888.

[2] J. E. Gray—"Spicilegia Zoologica." London. 1828.

is different from that which Duchassaing and Michelotti obtained at St. Thomas and Guadeloupe, relying probably on the discrepancies between the poor figures given by the latter authors and the more correct one which Gray has given. He consequently retains the specific term *tuberculatus* for Gray's form, proposing for Duchassaing and Michelotti's the name *Duchassaingi*. There is little room for doubt, however, that the two forms are identical; my observations have shown that the species is to be referred to the genus *Mammillifera* as defined by Erdmann.

The specimens from the Bermudas were either solitary, attached to a piece of rock by a base only very slightly expanded, or else were grouped together in twos or threes, in which case they were united by a slightly-developed, flat or slightly tubular cœnenchyma. In none had the cœnenchyma any such tubular or stolon-like form as is shown in the figure given by Duchassaing and Michelotti. Judging from the specimens I studied, the tendency to form a cœnenchyma is slight.

The polyps (Pl. 11, fig. 5) vary in height from 1.3–2.7 cm.; their diameter being about 0.7–0.9 cm. The column is marked by six or eight distinct annular grooves, and by from twenty to twenty-five longitudinal ones. In the lower part of the column the ridges formed by these longitudinal grooves are entire, but higher up they begin to be divided into series of tubercles, a row of these corresponding to each ridge. These tubercles increase in size towards the margin and several become grouped together upon elevations of the column wall, giving rise to mulberry-like protuberances. Near the margin the tubercles suddenly cease, forming, in contracted specimens, a strong ridge bounding the dome-shaped area which forms in such specimens the summit. This dome-shaped area belongs to the column, the animal being strongly contracted, and though without tubercles shows clearly the continuation upwards upon it of the longitudinal furrows, and is, accordingly, marked by a series of radiating ridges.

In structure the tubercles of the column are solid, being

elevations of the mesoglœa. This tissue throughout the
column is very thick, measuring on the average 1 mm. in
thickness. It presents numerous anastomosing canals filled
with cells, as well as the delicate canals which have been described by Erdmann and others, very distinctly. These canals
are without doubt processes from the large canals, and the
structure of the zoanthan mesoglœa may be compared to that
of a bone, such as a frog's femur, the anastomosing canals
being compared to the lacunæ, and the delicate canals to the
canaliculi. My preparations of *M. tuberculata* seem to show
that the lacunæ arise from both the ectoderm and the endoderm. In some of my sections deep bays can be seen running
from the endoderm up into the mesoglœa, and from their ends
and sides numerous canaliculi can be seen branching out.
These bays are found in various stages of enclosure by the
mesoglœa, the cells which they contain being in some cases
continuous with the general endoderm, in other cases almost
separated from it, and finally quite so. So too with the ectoderm. The lacunæ which have just been formed in this manner are much larger than the majority of those scattered
through the mesoglœa, these frequently consisting of only a
few cells or even a single cell, and further, the newly-formed
lacunæ usually contain zooxanthellæ, which are rare in the
older ones. It would seem as if many of the newly-formed
lacunæ become divided into smaller portions which separate
from each other, except by the delicate canaliculi, and at the
same time undergo an alteration in the histological structure
of their cells, the zooxanthellæ disappearing, and the cells becoming filled with refractive, deeply-staining granules. It
seems not improbable that these altered cells are concerned in
the formation of the mesoglœa, their granules being particles
which will later on be added to the matrix of the mesoglœa.

Upon the outside of the column is a thin cuticle (Pl. 11, fig.
6, cu) similar to what occurs in *Z. sociatus* and *Z. flos-marinus*.
Andres[*] considers this to be merely a differentiation or hard-.

---

[*] A. Andres. On a new genus and species of Zoanthina Malacodermata (*Pan
ceria spongiosa*, sp. n.)—Quart. Journ. Micros. Sci. N. S. Vol xvii. 1887.

ening of the external layers of the mesoglœa, but I cannot
agree with this view. It is a clearly defined layer external to
the mesoglœa, and appears quite different in composition and
behavior to staining fluids from that tissue. Below this cuticle
comes a layer of mesoglœa for which Andres's term subcuticula
may be employed. The distinction between the cuticle and
this layer has been overlooked by most authors. It was recognized by Kölliker[1], however, who believed it to be a portion
of the cuticle. Andres recognized its true nature, considering
it simply a continuation of the mesoglœa.

Below the subcuticula is the ectoderm (Pl. 11, fig. 6, ec),
which forms a layer 0.08 mm. in thickness. It is not continuous, however, but is divided into more or less cubical masses
by columns of mesoglœa extending from the general mass of
that tissue to the subcuticula. A peculiar feature of the ectoderm of this species is the presence in it of zooxanthellæ. In
adult actinians these structures are usually confined to the endoderm, but I have observed them in the ectoderm in free-
swimming larvæ, in which layer they also occur according to
H. V. Wilson[2] in the embryos of the coral *Manicina*. It is possible that their presence in the ectoderm of *M. tuberculata* is due
to the thick cuticle and subcuticula preventing a rapid aeration
of the ectoderm cells, and so, by favoring the accumulation to
a certain extent of carbon dioxide, producing favorable conditions for the growth of the parasitic algæ. The ectoderm thus
buried in the mesoglœa evidently corresponds with what
Kölliker, in the admirable account he has given of the zoanthan mesoglœa,[3] terms "eine zusammenhängende Schicht
drüsenartiger Körper" and which he believed to correspond to
the ectoderm.

The endoderm consists of low cells containing numerous zooxanthellæ. In the upper part of the column, extending from

---

[1] Kölliker. *Icones Histologicæ.* Leipzig. 1865.

[2] H. V. Wilson. On the Development of Manicina areolata. Journal of Morphology. Vol. II. 1888.

[3] A. Kölliker, *loc. cit.*

the margin to the upper row of tubercles, is a single strong sphincter muscle imbedded in the mesoglœa, and occupying nearly its whole thickness.

All the specimens were in a state of strong contraction, and I was not able to see the tentacles. Duchassaing and Michelotti state that they are small tubercles. My sections show that they are arranged in two cycles. It is also evident that they are short, but they can scarcely be termed tubercles. Their mesoglœa is thick, especially toward the base, thinning out somewhat towards the apex. Its outer surface is thrown into rather strong muscular processes.

The surface of the stomatodæum is thrown into numerous rather high folds, the ectoderm being elevated on slender processes of the mesoglœa.

The mesenteries are arranged on the microtypus and number twenty-two pairs. Towards their base the mesoglœa is very thick, diminishing gradually towards the distal edge. Just at the base there is a sudden diminution of the thickness, so that they are attached to the column wall by a thin pedicle. The basal portion contains the usual canal, and in addition there are numerous lacunæ, similar to those of the column wall in every respect. *M. tuberculata* is hermaphrodite, and I am able to add this particular to the definition of the genus given by Erdmann.* I could not make out any regularity in the arrangement of the reproductive elements on the different mesenteries, nor did there seem to be any definiteness in their position in any one mesentery. Sometimes a mesentery would possess ova only, but usually each one presented both ova and spermatozoa.

Corticifera ocellata (Ellis).

Alcyonium ocellatum. Ellis and Solander, 1786.

Palythoa ocellata. Lamouroux, 1821.

A number of small colonies of a *Corticifera* were obtained at Shelly Bay, and were accompanied by a label referring them

* A. Erdmann, *loc. cit.*

to the above species. The term *ocellata* was first given by Ellis and Solander to a form which, however, was very poorly characterized, so much so that certainty of identification is impossible. The only statement in the description of which use may be made is that the polyps are rust-colored. Later authors simply copied Ellis and Solander's description, until Dana,* evidently relying on the figure which accompanies the earlier description, adds the characteristic that the polyps, though imbedded in cœnenchyma throughout the greatest part of their extent, are yet free above. Duchassaing and Michelotti in their paper of 1860 described a form under this name which differs somewhat from the original type species, and is probably to be considered, as Andres has done, a distinct form. In their later paper they make this form identical with a form they name *Polythoa mammillosa*, a name taken from a second imperfectly characterized form mentioned by Ellis and Solander. In fact, so much confusion is introduced by Duchassaing and Michelotti as to render it very difficult, if not impossible, to ascertain what forms they are really describing.

Under the circumstances I have thought it well to retain the name which accompanied the specimens, and trust that the following description will sufficiently characterize them to allow of the identification in the future.

The polyps are grouped together in small masses, and project decidedly above the surface of the cœnenchyma. Their height measured from the lower surface of the cœnenchyma is 1-2 cm., and their breadth, measured at the summit, about 0·7 cm. in the fully grown individuals. The polyps and cœnenchyma are densely incrusted with particles of sand and other foreign bodies, and are of a grayish sandy color, sometimes deepening to a rust color.

Upon the outside of the column is a rather thick cuticle, but I was not able to discover whether or not a layer of mesoglœa intervened between this and the ectodermal cells.

---

* J. D. Dana.—Zoophytes. United States Exploring Expedition. 1849.

NATURAL ARCH, TUCKER'S TOWN.

The outer portion of the mesoglœa for about half its thickness has imbedded in it foreign bodies, and when decalcified is fenestrated by the numerous cavities previously occupied by them. The internal portion of the layer presents the structural features found in other Zoanthidæ, but it is to be noticed that foreign bodies occur in the so-called "nutritive canals" or lacunæ. The sphincter muscle is imbedded in the mesoglœa, is single, and consists of a single row of cavities containing muscle fibres.

The tentacles are arranged in two rows, and are apparently fifty-six in number in the specimens examined. Their outer muscular layer is weak, and the mesoglœa is homogeneous, except upon the outer face of the tentacles, where it contains a number of granular cells similar to those occurring in the column mesoglœa in this and other forms already described. Zooxanthellæ occur in the ectoderm.

The ectoderm of the disc is peculiar. It consists of high, much-vacuolated cells which contain, like the ectoderm of the tentacles, zooxanthellæ. I have found this peculiar structure of the disc ectoderm in no other Zoanthids. Unfortunately the preservation of the specimens was not sufficiently good to allow of the histological details being studied. The gonidial groove of the stomatodæum is rather broad, and the mesoglœa lining is thickened and truncated upon the endodermal side, the macrodirectives being inserted into each angle of the truncation.

The mesenteries are arranged on the microtypus, there being about twenty-six pairs. The basal canal is large, and contains foreign particles similar to those found in the lacunæ of the column. The mesoglœa is thickened towards the base of the mesenteries and contains, in addition to the basal canal, several others nearly circular in section and completely filled with spherical granular cells. The endoderm throughout contains zooxanthellæ. No reproductive organs were present.

Corticifera glareola, Les.

 Corticifera glareola. Lesueur. 1817.

 Palythoa glareola. Milne-Edwards. 1857.

The identification of this form depends mainly on the coloration, which Professor Heilprin informs me is sufficiently similar to Lesueur's description.

The polyps form encrusting masses, and are so deeply imbedded in the cœnenchyma, that in contraction a slight depression alone indicates the position of the various individuals, or in some cases a slight annular elevation. The species is by this peculiarity readily distinguishable from *C. ocellata*, as well as from *C. flava* of the Bahamas, which stands in an intermediate position as far as the projection of the polyps above the cœnenchyma is concerned. The form described from the Bermudas by Erdmann, and named *C. lutea* by Hertwig, resembles *C. glareola* in this respect, but appears to differ from it in other points.

The mesoglœa is, with the exception of a narrow band immediately adjoining the endoderm of the polyps, richly supplied with imbedded foreign bodies, so that the entire colony is very hard, almost stony in its consistency. *C. ocellata* is much less richly provided with foreign particles, and the same is the case with Hertwig's *C. lutea*. Whether this is a characteristic of sufficient importance for specific distinction can only be ascertained by the examination of numerous specimens of some species, obtained from different localities and living under different conditions. In fact, our knowledge of the histology of the Zoanthidæ is not yet sufficiently advanced to enable us to ascertain what features are of systematic importance and what are liable to extensive individual variation.

The sphincter muscle resembles closely that of Hertwig's *C. lutea*. It is imbedded in the mesoglœa and is single, consisting of a single row of cavities which are entirely confined to the portion of the column which is invaginated during contraction. All the cavities contain muscle cells, and there are none of the empty spaces with clearly defined walls such as occur in *C. flava*.

The mesenteries are arranged on the microtypus, and in the specimens examined there were about eighteen pairs only.

The mesoglœa is delicate, and is not dilated towards the base as in *C. ocellata*, and in consequence, the basal canal is elongated. Notwithstanding that the specimens were very much macerated it was possible to perceive that a digestive area, similar to that described as occurring in *Z. flos-marinus*, was present, just below the stomatodæum. No reproductive organs were present.

The stomatodæum presented the pyriform, truncated shape which has been described for other members of the genus.

It seems not improbable that the form described by Hertwig as *C. lutea* may be identical with this. Alcoholic specimens of *C. glareola* show no trace of the coloration of the living forms, but are of a universal sandy color. In the very slight prominence of the polyps above the cœnenchyma, in the structure of the sphincter muscle, and in the slenderness of the mesenteries there is agreement between the two, and these are points which will probably prove to be of systematic importance. On the other hand, there is dissimilarity in the extent of the incrustation by foreign bodies, in the pigmentation of the endoderm, which is wanting in *C. glareola*, and apparently in the extent of the development of the longitudinal muscles of the mesenteries, which cannot be said to be well developed in *C. glareola*. This last character is probably of importance, but the first two are probably subject to variation depending upon the conditions of life and the food.

The evidence then, seems to be in favor of the identity of the two forms, in which case the name here used has the priority. It seems to me very doubtful, indeed, if Hertwig's identification of the Bermuda form with Quoy and Gaimard's *C. lutea* from the Feejee islands is correct. The only point of correspondence, judging from the description and figures given by Quoy and Gaimard,[*] is the slight prominence of the polyps above the cœnenchyma when in contraction.

Gemmaria Rusei, Duch. and Mich. (Pl. 11, figs. 7–9.)

Gemmaria Rusei.   Duchassaing and Michelotti. 1860.

----

[*] Quoy and Gaimard, *Zoologie du Voyage de la Corvette l'Astrolabe.* Paris. 1833.

I was pleased to find in the Bermuda collection several specimens of a form which evidently belongs to the same genus as the form from the Bahamas which I described as *Gemmaria isolata*. Several anatomical features are common to the two. and I am now able to give other characteristics which may serve to distinguish the genus more definitely than was done in my former paper.

The polyps of *G. Rusei* (Pl. 11, fig. 7) are solitary, being attached to pebbles without the development of any coenenchyma. The specimens were obtained at North Rock, and are five in number. The upper portion of the column is larger than the lower, so that the polyps have the shape of a short stout club; the lower portion is transversely wrinkled even in the expanded condition, as is noted in the label accompanying the specimens. The height of the column is about 2·5 cm. in the largest specimens; the diameter of the upper part is 0·65 cm. and of the lower 0·5 cm. The color is stated on the label to have been "cinereous throughout."

The column wall is rather thin, and is occupied throughout nearly its entire thickness by foreign bodies. The ectoderm is covered externally by a cuticle, but I was unable to ascertain whether a layer of mesoglœa intervened between this and the surface of the ectoderm. The structure of the thin layer of mesoglœa unoccupied by foreign bodies is as in other Zoanthidæ, and calls for no special comment. The sphincter is single, and imbedded in the mesoglœa; it consists for the most part of a single layer of cavities, but thickens somewhat towards its upper end. All the cavities contain muscle cells, there being none of the empty cavities described in *G. isolata*.

The tentacles are arranged in two cycles, and have only a very weak ectodermal musculature, as is also the case in *G. isolata*. Towards the base and upon the outer surface the mesoglœa contains peculiar granular cells, and occasionally enclosures of foreign bodies, and this likewise occurs in *G. isolata*.

The disc is traversed by a number of ridges which radiate from the peristome to the margin, a ridge corresponding to each tentacle of the outer cycle. The elevations are produced by thickenings of the mesoglœa (Pl. 11, fig. 9), and along each ridge the ectodermal muscle cells are more numerous and larger than elsewhere. *G. isolata* presents similar structures. Zooxanthellæ occur in the ectoderm of the disc, and tentacles in both forms. The enclosures in the mesoglœa of the disc, which I thought might possibly be muscle cells in *G. isolata*, are seen in *G. Rusei* to be comparable to the lacunæ of the column wall.

The mesoglœa of the stomatodæum in both species of *Gemmaria* has enclosures of granular cells (Pl. 11, fig. 8), as a rule one such enclosure opposite the insertion of each mesentery, especially in the upper part of the stomatodæum, the arrangement being lost in the lower part. The gonidial groove has the same shape as that of *G. isolata*.

The mesenteries are arranged in thirty-one pairs and are on the microtypus. The mesoglœa thickens towards the base so that the basal canal is almost circular and not elongated as in *G. isolata*. No reproductive organs were present.

The description given by Duchassaing and Michelotti of *Gemmaria Rusei*, with which I identify this form, is very imperfect, but so far as it goes it applies to the Bermuda species. The form described by Gray* as *Triga Philippinensis* is very similar in external form and is in all probability a *Gemmaria*.

Of the forms described above, no less than seven, viz: *Condylactis passiflora, Phymanthus crucifer, Zoanthus flos-marinus, Mammillifera turberculata, Corticifera ocellata, C. glareola* and *Gemmaria Rusei*, are represented in the West Indian fauna, and of the other three, the genera *Aiptasia* and *Phyllactis* also occur in the islands to the South, leaving only the genus *Diplactis* as a characteristic form of the Bermudas. No doubt a systematic search for actinians in the Bermudas would lead to the discovery of a greater number of West Indian forms, but the proportion of common forms given above is sufficient

---

* J. E. Gray. Notes on Zoanthinæ with Descriptions of some New Genera. Proc. Zool. Soc. London, 1867.

to show that the actinian fauna of the Bermudas has been derived from that of the West Indies.

EXPLANATION OF PLATES.

bc.= basal canal.

c.=column wall.

cm.=circular muscles.

cu.=cuticle.

d.=disc.

di.=digestive region of mesenterial filament.

Ec.=Ectoderm.

en.=endoderm.

fr.=fronds.

m.=margin.

p.=tentaculiform fronds.

sp.=sphincter.

sp¹.=lower sphincter.

t.=tentacle.

PLATE 10.

1. Transverse section through the middle region of the sphincter of *Aiptasia* sp. (?) × 350.

2. Longitudinal section through the upper half of the column wall of *Aiptasia* sp. (?) × 40.

3. Transverse section through the outer edge of the longitudinal mesenterial muscles of a specimen of *Condylactis passiflora* from the Bermudas. × 42.

4. Longitudinal section through the margin and adjacent parts of *Diplactis Bermudensis.* × 24.

5. Longitudinal section through the margin and adjacent parts of *Oulactis fasciculata.* × 21.

6. Transverse section through the longitudinal mesenterial muscles below the stomatodæum in *Diplactis Bermudensis.* × 40.

PLATE 11.

1. Perfect mesentery of *Diplactis Bermudensis.* Natural size.

2. Portion of transverse section of sphincter of *Diplactis Bermudensis.* × 100.

3. Transverse section of mesenterial filament of *Zoanthus flos-marinus* just below the stomatodæum. × 120.

4. Transverse section of perfect mesentery of *Zoanthus flos-marinus* slightly below the stomatodæum. × 50.

5. *Mammillifera tuberculata.* Natural size.

6. One-fourth of a portion of a longitudinal section through the column wall of *M. tuberculata.* × 200.

7. *Gemmaria Rusei.* Natural size.

8. Transverse section through the gonidial groove of *Gemmaria Rusei.* × 65.

9. Transverse section through upper part of column of *Gemmaria Rusei.* × 24.

HYDROID-CORALS.

Apparently both of the common West India species of millepore, *Millepora alcicornis* and *M. filiformis,* are found in the Bermudas; at any rate, forms answering to these are found in our collections. I feel doubtful, however, if the two should not properly be classed as a single species, seeing how great is the individual variation, and how closely the species approximate one another. It is certainly not easy to separate them by the characters which have been generally indicated by systematists.

VII.

THE ZOOLOGY OF THE BERMUDAS (*continued*).

HOLOTHURIA.

The animals of this order are in places exceedingly abundant; indeed, excepting the corals, they may be said to constitute the most distinctive feature of the fauna of the sand bottoms. Where other forms are apparently entirely absent, the black masses of the great Stichopus stand out in prominent relief over the white bottom. Motionless, seemingly, during the greater part of their existence, these singular creatures present the appearance of big black blotches on the sand, of which they consume, whether for nourishment or otherwise, vast quantities. All the individuals that were opened had their digestive tracts completely filled with calcareous particles.

The following are the species of holothurians observed by us, only one of which, I believe, had hitherto been noted from the Bermudas:

Holothuria Floridana, Pourtalès. (**Holothuria atra**, Jäger.) Pl. 12, figs. 6, 6a, 7, 7a.

I identify with this species five small individuals of an olive-green color which were obtained in Castle Harbor, and which in a general way agree with the description of the species given by Pourtalès (Proc. American Assoc., 1851, p. 12). Unfortunately, no figure accompanies the description, and that part which pertains to the calcareous bodies embodied in the skin is too vague to permit of specific determination. Selenka (*Zeitschrift für wissenschaftliche Zoologie*, xvii, p. 324, 1867) has supplemented the original description with further details of structure and with illustrations of the spicules, which practically leave no doubt in my mind that the Bermudian forms, even though differing somewhat from the type described by Pourtalès, are really that species. I have examined the spicular bodies of all the individuals, and find that they exhibit considerable variation (Pl. , figs. 6, 6*a*, 7, 7*a*). This is especially noticeable in the form of the stools. I really doubt if very much dependence can be placed upon these bodies as furnishing characters for specific distinction. I also find a certain amount of variation in the number of tentacles. Thus, while four of the individuals have the normal number of tentacles, 20, one has only 10, although in all other essentials of structure it agrees with the remaining four. The dorsal surface is distinctly papillate. The elongated yellowish pedicels of the ventral surface are irregularly distributed, as stated by Selenka, and I could not determine any strictly linear disposition such as is indicated by Pourtalès.

The largest specimen measures about two and a half inches.

Semper, Ludwig, and Lampert (*Die Seewalzen*, Semper's *Reisen im Archipel der Philippinen*, 1885, p. 86) identify this species with the *Holothuria atra* of Jäger (1833), whose range is made to be practically cosmopolitan—extending from the Radack Archipelago and the Sandwich Islands to Adelaide, Zanzibar, the Red Sea, and the West Indies—but on this point I can offer no satisfactory evidence, never having had an opportunity to examine authentic specimens of Jäger's species.

Holothuria captiva, Ludwig. (Pl. 12, figs. 4, 4a.)

Two individuals, agreeing with the species described by Ludwig from the Barbados.

Holothuria abbreviata, n. sp. (Pl. 12, figs. 5, 8, 8a.)

Among the smaller forms of holothurians is one which in many of its characters agrees most closely with Ludwig's *H. captiva*, but yet differs to such an extent as to compel me to recognize it as a distinct species. Indeed, by many systematists it would probably be made the type of a distinct sub-genus or genus. The distinguishing peculiarity is the abrupt truncation of the body, which carries the vent on the dorsal surface, immediately about the extremital border. In the single specimen before me I could determine only 17 tentacles, with as many tentacular vesicles, and but a single Polian body. A large Cuvierian bundle is present. The pedicels are arranged ventrally in three more or less distinct rows. Color olive green. Length about two inches.

The stools, buttons, and fenestrated plates of the pedicels are figured on plate 12. It will be seen that in general they bear a close resemblance to those of *Holothuria captiva*, but the rounded summits of the stools serve readily to distinguish them from the somewhat similar, but more strictly castellated, bodies of the other species.

SEMPERIA.

Semperia Bermudensis, n. sp.　(Pl. 12, figs. 2, 2a, 3.)

Body cylindrical, spindle-shaped, tapering almost equally to both extremities. Tentacles 10, of which four are shorter than the remaining 6; pedicels crowded, arranged in five broad rows, and scattered over the interambulacral areas; two genital bundles, with very numerous non-divided, and greatly elongated filaments; two Polian vesicles; two long respiratory trees. Color grayish white, minutely speckled with brown; five narrow longitudinal brown bands separating the ambulacral areas. Length about $3\frac{1}{2}$ inches.

Calcareous bodies consisting of baskets, knotted and smooth buttons, and perforated more or less circular disks; pedicels with fenestrated plates. Calcareous ring with long back processes for the attachment of the powerful retractor muscles.

One specimen, from the north shore about a half-mile west of Flatts Village.

I first mistook this species for the *Semperia* (*Colochirus*) *gemmata* of Pourtalès (Proc. Amer. Assoc. 1851, p. 11), described from Sullivan's Island, coast of South Carolina, but the more exact descriptions and figures of that species given by Selenka and Lampert convince me that it is quite distinct. Both species are of a grayish-white color, but no mention is made by either of the authors above quoted of the existence in the Carolinian form of the five longitudinal brown bands which extend over the entire length of the Bermudian species. Apart from this, *Semperia Bermudensis* differs in the disposition of the tentacles, the greater number of Polian vesicles, and the character of the spicular buttons, which are in the greater number of instances strongly knotted. The posterior processes of the calcareous ring appear also to be much more elongated.

From *Semperia* (*Cucumaria*) *punctata*, described by Ludwig from the Barbados (*Arbeiten aus dem zoolog. zootom. Instituts in Würzburg*, ii, 1875, p. 82) the species differs, apart from the general scheme of coloring—tentacles as well as body—in the different disposition of the tentacles (9 equal in *S. punctata*, according to Ludwig), the smaller number of Polian vesicles (5 in *S. punctata*), and in the much greater number of filaments composing the genital bundles. The vent does not appear to have been rayed.

Ludwig states that there are in his species no calcareous teeth about the anal aperture, whereas Lampert just as positively asserts that they are present (Semper, *Philippinen*, 1885, p. 152). None such were detected in the Bermudian form.

STICHOPUS.

Stichopus diaboli, n. sp.   (Pl. 13, figs. 1, 1a, 1b, 2.)

Body stout, more or less quadrangular, flattened ventrally, and bearing two rows of prominent marginal wart-like, tubercles; sometimes two additional rows of minor tubercles are noticeable on the lateral margins of the dorsum. Tentacles 20, unequal. Dorsal papillæ scattered, not prominent, leaving the surface nearly smooth. Pedicels and papillæ on ventral surface arranged in three broad bands, which are more or less distinct for the entire length of the body, but most distinct near the extremities; numerous in each transverse row.

The body-cavity is largely occupied by the greatly developed, and finely dissected, respiratory apparatus, and by the loops of the variously branched genital organs, which are disposed in two great bundles. Tentacular vesicles present. Two Polian vesicles. Calcareous ring with long back processes.

Calcareous bodies in the form of stools very numerous (Pl. 13, fig. 16). C-shaped bodies very scanty, and possibly in some cases entirely wanting.

Color black, somewhat more intensely so on the dorsal surface, becoming Vandyke brown or chocolate in alcohol.

Length, about one foot; width of corresponding animal about three inches.

Abundant over the sandy floor of the entrance to Harrington Sound, opposite Flatts Village, in Harrington Sound, and in Castle Harbor, whence it was obtained in several of our dredgings.

I have little doubt that this species is the dark-brown form which is referred to by Théel as having been obtained by the officers of the Challenger at the Bermudas, and which is doubtfully referred by that authority to Semper's *Stichopus Haytiensis* (Report on the Holothuroidea, Challenger Reports, Zoology, XIV, p. 162). Only a single specimen appears to have been obtained, which when examined was too deformed to permit of positive specific determination. I cannot agree with Théel's determination. Apart from the differences which Théel himself points out, is the great difference in coloring. Semper (*Reisen, Philippinen, Holothurien*, 1868, p. 75) states that his species is dark chocolate-brown, blotched with yellow spots, which form five longitudinal bands, corresponding to the interradii. No such coloration is visible in our species, although probably we observed as many as a hundred individuals, all of which were uniformly black. Semper's description of the coloring of *Stichopus Haytiensis* is re-stated by Lampert.

Stichopus xanthomela, n. sp.   (Pl. 12, fig. 1 ; Pl. 13, fig. 3.)

Body stout, flattened ventrally, and bearing on the basal margin two rows (one row on each side, as in the preceding

species) of prominent wart-like processes. Tentacles 18, unequal, whitish or gray, edged with brown. Dorsal papillæ
fairly prominent, scattered. Pedicels on ventral surface
crowded, arranged in three longitudinal series, five to eight, or
more, in each transverse row.

Body-cavity, as in the preceding, largely occupied by the
respiratory tree and the double genital bundle, the filamental
processes of the latter much finer than in *S. diaboli.* Tentacular vesicles present. One (?) Polian vesicle.

Calcareous bodies, in the form of stools (Pl. 13, fig. 3), very
numerous C-shaped bodies scarce, in the form of broadly-
opened calipers. Ground-color reddish-yellow, irregularly
blotched with black or very dark brown. The spots on the
ventral surface more or less coalescent in the median line,
forming there a broad longitudinal band, or entirely united to
form a uniformly dark-colored base; on the back, united into
two irregularly ramifying or wandering bands.

Length of longest specimen about ten inches; width about
two and a-half or three inches.

The same habitat as that of the preceding species, although
apparently much less abundant.

I strongly suspect that this is the form which Théel, in his
report on the Challenger holothurians (*loc. cit.,* p. 159), identifies with *Stichopus Möbii* (Semper), one specimen of which,
" rather deformed and compressed " when examined by Théel,
was obtained at the Bermudas. I assume the identity in this
case, as well as in that of the preceding species, on the ground
that the two species here described are the characteristic forms
of the archipelago, and it is barely possible that they could
have escaped the attention of the Challenger people. But the
identification with Semper's species appears to be erroneous. The resemblance to *Stichopus Möbii* appears to rest
talmos wholly upon the form of the spicules, which are largely
similar in many very distinct forms of *Stichopus,* and in a
general scheme of coloring. But Semper distinctly states
(*Holothurien, loc. cit.,* p. 246) that the characteristic spots are

almost wholly wanting on the ventral surface, and no mention
is made of their occurrence there by Lampert in his revision
of the species of the genus (*op. cit.*, p. 108). Moreover, Semper
affirms that the body is devoid of wart-like tubercles, whereas
such are quite prominent in the Bermudian form, although
not as prominent as in *Stichopus diaboli*. Théel, however,
makes no mention of the occurrence of tubercles in his single
specimen, but probably through contraction in alcohol their
existence had been effaced. The number of pedicels in each
transverse row seems also to be much more numerous in the
Bermudian species than in *Stichopus Möbii*.

Another apparently related form is *Stichopus errans* of
Ludwig (*Arbeiten zoolog. zootom. Inst.*, *Würzburg*, 1875, p. 97),
described from a specimen in the Hamburg Museum, reputed
to have come from the Barbados. But in this species there
appear likewise to be no lateral tubercles, nor is the coloring
like that of our species, although in this regard there may be
considerable variation. The number of tentacles is stated by
Ludwig to be 19, and their color yellow. The form from the
Barbados which is somewhat doubtfully referred by Théel (*loc.
cit.*, p. 191) to Ludwig's *S. errans* would seem to be more nearly
related to the Bermudian species.

ASTEROIDEA.

We obtained but a single species of star-fish on the Bermudian coast. This is the *Asterias Atlantica* of Verrill, a form
which had already been previously noted from the Bermudas
(Trans. Conn. Acad. Sciences, i, p. 368), and whose range
extends to the Abrolhos Reef, Brazil. With very few exceptions the rays were either six or eight in number, and of the
total number of individuals examined I believe that not over
two had five arms. The species exhibits a marked want of
constancy in ornamentation and coloring, the dorsal spines
being in some cases acute, while in others they are terminated
by a minute bead; again, while the maculation is brown in

some individuals, in others it is blue, or of both colors combined.

Asterias Atlantica, Verrill.

Common in the entrance to Harrington Sound, opposite Flatts Village—under stones; dredged in Harrington Sound.

Linckia Guildingii, Gray.

A single specimen, marked as having been collected by Mr. Janney in the Bermudas, is in the possession of the Academy of Natural Sciences.

OPHIUROIDEA.

Six species of ophiurians were obtained in our dredgings and under rock shelters, the greater number of which, so far as I am aware, had not hitherto been reported from the Bermudas. For a critical examination and review of the species I am indebted mainly to my assistant, Mr. J. E. Ives, who has made a careful study of all the species in the collections of the Academy of Natural Sciences. From an examination of many of these forms I feel satisfied that too much dependence should not be placed upon the constancy in minute details of either the form or relative size of the arm plates and their appendages, nor upon an exact scheme of coloration. These characters, and others that may be added, which have been drawn in very close limits by Mr. Lyman in his several memoirs, vary materially within the limits of the same individual, and render the discrimination of species which have been most "elaborately " defined as to exact lengths and breadths of the arm-shields and oval plates, the precise form and number of the arm-spines, etc., a matter of almost hopeless impossibility.

Ophiocoma crassispina, Say.

One specimen, taken at low water from the North Rock, which agrees perfectly with the species described by Say from the coast of Florida (Journ. Acad. Nat. Sci., Phila. v, p. 147). This species is generally considered to be identical with the *Ophiocoma (Ophiura) echinata* of Lamarck, but I am disposed to consider this identification erroneous, unless, indeed, several distinct forms, as has been averred by Müller and Troschel (*System der Asteriden*, 1842, p. 98), were included by Lamarck in his species. Two distinct forms, closely related to each other, certainly do occur in the West Indies, one of which, with more blunt arm spines, is clearly Say's species, while the other, with more elongated arm spines, and much less stoutly developed uppermost spine, more nearly corresponds to the general type of Lamarck's species.

Ophiocoma pumila, Lütken.

A fragmentary specimen; exact locality unknown. This species had been recorded by the Challenger from Bermuda.

Ophiostigma isacantha, Say.

Two very young specimens, dredged in Harrington Sound.

Ophiactis Mülleri, Lütken.

O. Krebsii, Lütken ?

Two very young specimens, dredged on the north shore between Bailey's Bay and Shelly Bay.

Ophionereis reticulata, Lütken.

Very abundant at low tide in the rock shelters of Shelly Bay; also under stones at the entrance to Harrington Sound.

Ophiomyxa flaccida, Lütken.

One specimen, dredged in Bailey's Bay.

ECHINOIDEA.

The number of species of echinoids observed by us is six, of which five had already previously been ascribed to the archipelago; *Cidaris tribuloides*, so far as I am aware, had not hitherto been collected—at any rate, I have been unable to find any mention of its occurrence there. One species, *Mellita sexforis*, we did not ourselves collect, the specimens in our possession having been kindly donated to us by local collectors.

Cidaris tribuloides, Bl.

Fairly abundant among the coral shelters of the North Rock.

Diadema setosa, Gray.

THE LANDING.

This species, one of the gems among sea-urchins, is exceedingly abundant in the flats about the North Rock. All the individuals occupied recesses in the coral growth, which they had by some means probably managed to keep open. It seems hardly likely that they should have crept into these shelters after they had been already formed, and that the association is one of mere selection.

The species is also abundant in the moderately deep water that lies within the reef border.

Hipponoë esculenta, Leske.

North Rock, and the deeper water within the growing reef.

Echinometra subangularis, Leske.

Several specimens from the flats about the North Rock. There is a certain amount of variation in the coloration of the spines which ranges from olive or sea-green to purple.

Toxopneustes variegatus, Lamk.

We found this species very abundant in Harrington Sound, where it rarely escaped being hauled up in our dredge. It seems to frequent the calcareous bottom to a depth of 10–12 fathoms, or even more. Probably the species is equally abundant elsewhere.

Mellita sexforis, Agassiz.

As before remarked, we did not ourselves obtain any specimens of this species. It is said to be abundant along the calcareous bottoms of some of the inlets, as, for example, opposite Flatts Village.

VIII.

ZOOLOGY OF THE BERMUDAS (continued).

CRUSTACEA.

For the following notes on the Crustacea I am principally
indebted to Mr. Witmer Stone, one of my assistants on the
trip, who has made a careful study of all the specimens, as well
as of the allied and identical species contained in the collections
of the Academy of Natural Sciences. In the case of in any
way doubtful forms I have personally satisfied myself as to
the determinations, particularly in cases where the geographical range appeared to indicate possible or probable
error. The occurrence in the Bermudas of a number of what
had hitherto been considered to be distinctively Pacific or Old
World types, as for example, *Palæmonella tenuipes* (Sooloo Sea),
Palemon affinis (Pacific), *Penæus retutinus* (Pacific)—may be
considered positive, even though it be opposed to the common
facts of zoogeography. But this anomaly in distribution is
again repeated among the Mollusca, as will be seen in the
enumeration of species further on.

The total number of species here enumerated is not very
large, but yet it is considerably in excess of the number
published in any previous paper, probably one-half of the
species being now for the first time credited to the Bermudas.
The species of some of the remaining groups—the Isopoda,
Amphipoda—still await analysis and determination.

BRACHYURA.

Microphrys bicornutus, Latr.

Three females and one male, collected on the beach at the entrance to Harrington Sound.

Mithraculus hirsutipes, Kingsley.

Two males and one small female, which agree in every way with the description of the species given by Kingsley (Proc. Bost. Soc. Nat. Hist., 20, p. 147), except in the number of teeth on the fingers, a character which appears to be very variable. The three individuals differ in this respect among themselves.

Actæa setigera, Milne-Edwards.

One male dredged off Shelly Bay. The individual differs from the description given by Milne-Edwards (*Nouv. Arch. du Mus. d' Hist. Nat.* i, p. 271, pl. xviii, fig. 2) in having the color of the outside of the hands red, instead of black. It however agrees precisely with specimens attributed to Milne-Edwards' species in the collections of the Academy, and labeled as coming from the Florida reefs. The species has also been recorded from Cuba

Panopæus Herbstii, var. serrata, De Saussen.

Numerous small specimens, both male and female, from under stones on the beach of St. George's Causeway, and at the mouth of Harrington Sound. The specimens vary greatly in color, some being very light, others dark brown, while a few are reddish ; otherwise they are identical in structure.

The species, described in the *Hist. Nat. du Mexique et des Antilles* (*Crustac.*, p. 16, pl. 1, fig. 7), had previously been recorded from the Bermudas.

Lobopilumnis Agassizii, Stimpson.

One small male, agreeing well with Stimpson's description (Bull. Mus. Comp. Zool., ii, p. 142) except in that it lacks the subhepatic spine. Recorded from Bermuda and Florida.

Neptunus hastatus, L.

N. dicanthus.

Two small males.

Geocarcinus lateralis, Frem.

Numerous large specimens, from the banks and fields near the south shore. We found them specially abundant near the locality known as Spanish Mark or the Chequer Board, and again not far from Peniston Pond. The burrows in places extend diagonally three or four feet, or even more, beneath the surface, and the animals, rapidly retreating into these, are frequently difficult to capture.

This is, doubtless, the species that is referred to by Willemoës-Suhm in the Challenger narrative as *Gecarcinus lateralis*, and is apparently the *G. lagostoma* (?) described by Miers in the systematic portion of the Challenger Reports (Zoology, XVII, p. 218), in so far as this description applies to the single Bermuda specimen.

Nautilograpsus minutus, L.

One small specimen dredged off Shelly Bay.

Grapsus maculatus, Catesby.

One large female, and numerous empty shells from Harris's Bay, south shore.

Pachygrapsus transversus, Gibbes.

Numerous specimens, including ovigerous females; very abundant on the rocks about the mouth of Harrington Sound, and also on the Pigeon Rocks, Bailey's Bay.

Recorded from Florida, West Indies, Australia.

Cyclograpsus integer, Milne-Edwards.

One small female. Species recorded from Brazil and Florida.

Goniopsis cruentatus, Latr.

One female, from the mangrove swamp of Hungary Bay, south shore. Although the species was very abundant at this locality we only succeeded in catching a single individual. The mangrove crab, or "mangrove climber," as the animal is sometimes called, burrows among the thickets of mangrove stems and roots, up which it not infrequently climbs to a height of several feet. The great similarity existing between its coloring and that of the bright and partially withered leaves of the mangrove, especially in the shades of yellow and red, renders the animal difficult of detection, and often at a distance of only a few feet, buried among the fallen leaves, these agile creatures escaped observation, even when attentively sought after. We have here one of the most remarkable instances of protective coloring, or semi-mimicry, with which I am acquainted.

Sesarme cinerea, Bosc.

Numerous specimens, from the beach of Flatts Village. The species was seen almost everywhere scampering over the rocks.

Calappa flammea, Herbst.

A single male individual obtained through purchase. Species previously recorded from the Bermudas.

ANOMURA.

Petrolisthes armata, Gibbes.

Five specimens, obtained on the beach of Flatts Village, which appear to be identical with the form described under this name from Florida (Proc. Amer. Assoc., 1850, p. 190).

Cenobita Diogenes, Latr.

A number of living specimens obtained at Wistowe, opposite Flatts Village, and kindly presented to us by Miss Edith Allen, daughter of the American Consul. Most of the animals are still living (July), and apparently flourishing, twelve months after their capture. The shells occupied by the largest individuals are those of *Natica catenoides*.

Calcinus obscurus, Stimpson.

Several specimens obtained on the beach of Flatts Village.

Clibenarius (Pagurus) tricolor, Gibbes.

Numerous on the beach of Flatts Village and at the St. George's Causeway ; under stones, etc.

MACRURA.

Palinurus Americanus, Lamk.

We observed a number of specimens of the large Bermuda crayfish, but unfortunately obtained none. I am unable, therefore, to state positively if the species is correctly referred, but in all probability it is the same as the common West Indian form.

Scyllarus sculptus, Milne-Edwards.

One specimen, purchased at the Crawl, which agrees with Milne-Edwards' description (*Hist. Nat. des Crust.*, ii, p. 283) and Lamarck's illustration in the *Encyclopédie*, pl. 320. The locality of the original specimen appears to have been unknown, nor have I been able to obtain data regarding this species from any of the later writers, by many of whom it is entirely ignored.

Alpheus avarus, Fabr.

A. *Edwardsii*, Audouin.

A. *Bermudensis*, Spence Bate.

A series of some twenty specimens collected at the same locality shows considerable variety of form. The smaller specimens are evidently the *A. Bermudensis* of the Challenger Reports, while the larger ones, agreeing with these in the structure of the head, etc., more nearly approximate in the configuration of the hand *A. avarus* and *A. Edwardsii*, the former a common Old World species, and the latter, a species described from the Cape Verde Islands. Our series contains what might be considered undoubted representatives of all three (so-called) species, showing all the gradations that unite, or separate the forms from one another. Hence, I am constrained to look upon them as mere varietal forms of a single species, the *Alpheus avarus* of Fabricius. The older the specimens, the more deeply grooved is in most cases the hand.

Alpheus minus, Say.

A number of species taken from sponges and tunicates collected in Harrington Sound. All the individuals were of small size, measuring rather less than an inch in length, although the females were abundantly provided with eggs.

Alpheus formosus, Gibbes.

One specimen (dredged) which agrees well with Gibbes' description (Proc. Amer. Assoc., 1850, p. 196), and seems to indicate that the species is distinct from *Alpheus minus*, with which it is united by Kingsley. The specimen is larger than any of the individuals of *A. minus*, and is also differently colored, although appearing identical in alcohol.

Palæmonella tenuipes, Dana.

Several specimens dredged off Shelly Bay, which agree perfectly with the species described by Dana from the Sooloo Sea (U. S. Exploring Expedition, Crustacea, p. 582). The remarkable distribution here indicated induced me to make a very careful examination of the Bermudian species, which has left no doubt in my mind as to the identity of the forms from the antipodal regions of the earth's surface. The only other known species of Palæmonella, *P. orientalis* (Dana), is likewise an inhabitant of the Sooloo Sea (Dana, *op. cit.;* Spence Bate, Challenger Reports, Zoology, XXIV, p. 786).

Palæmon affinis, Milne-Edwards.

Numerous specimens from shallow water, Castle Harbor. All are exactly like one another except in the number of teeth on the beak, which may be 8 above and 4 below, or in relations of 8–3, 7–3, 9–3, and 9–4. This character is manifestly a very variable one, and, therefore, of little or no value from a classificatory point of view. The specimens agree well with the descriptions and figures of *A. affinis*, although that species has hitherto been recorded, as far as I am aware, only from the Pacific (obtained by Dana off New Zealand). The species is near to the Eurafrican *P. squilla*, but yet sufficiently distinct to permit of ready recognition as only an allied form.

It is remarkable, in view of the distribution and the number of specimens that we obtained of this species, and the position of the island group, that we should have failed to obtain any individuals of the common form of the eastern United States, *Palæmon vulgaris*. Whether the species is entirely absent or not I cannot of course say, but it is surprising that it should not have been observed by us.

Penæus velutinus. Dana.

One specimen, which agrees with the figure and description of the species obtained by Dana off the Sandwich Islands (U. S. Exploring Expedition, Crustacea, p. 604), and which was subsequently collected by the Challenger party at various points in the Pacific, and between Australia and New Guinea (Challenger Reports, Zoology, XXIV, p. 253). This species, as well as all the immediately related forms, has, as far as I know, been found thus far only in the Pacific. The case is, therefore, another example of remarkable geographical distribution.

STOMATOPODA.

Gonodactylus chiragra, Latr.

One specimen from the beach of Flatts Village.

OBSERVATIONS ON THE INSECTS OF THE BERMUDAS.

BY

P. R. UHLER.

The present list of insects enumerates chiefly those brought together by the recent exploration of Prof. Heilprin, and it does not include the specimens belonging to the orders Coleoptera, Lepidoptera, and Hymenoptera. Although constituting only a small collection, it is of much interest as throwing new light upon a recently constituted fauna which has been only superficially noticed. But very few insects have hitherto been recorded from this group of coral islands, and much arduous collecting is still needed to gather a full series of the insects settled there. Representatives of large groups in nearly all the orders have not yet been reported as occurring on these islands, although we know that the conditions are favorable for the settlement and increase of many of them. As a notable instance we may cite the absence of such families as *Hydrobatidae, Notonectidae* and *Corisidae*, in the aquatic Hemiptera; and of the *Ephemeridae* among the Pseudoneuroptera. White ants and *Psocidae* likewise remain unrecorded; and the Diptera, a numerous host, seem to have been almost totally ignored. That part of the assemblage to which attention has been hitherto directed is almost entirely Nearctic in character, and corresponds with the fauna which exists in the eastern part of the United States from Cape Cod to northern Florida. A very few species, such as *Blabera Americana* and *Labidura riparia*, occur in the Bermudas, but they are wanderers which frequent vessels, and are liable to be transported to places where they make no permanent stay. There are, however, multitudes of Neotropical forms, residents of the West Indies and southern Florida, which we look for in connection with the palmettos and tropical fruit trees and shrubs that are now permanently settled in those islands; but these forms are still lacking in our collections. Can it be that these insect absentees are only such as live in the upper parts of the high trees, and that do not descend during the daylight so as to be noticed by collectors? Mr. J. M. Jones, in his "Naturalist in Bermuda," has given some account of a few insects belonging to this locality, but his attention seems to have been directed almost exclusively to the showy or more conspicuous kinds. It is therefore with earnest solicitude that we await the time when some acute collector will undertake to solve the problem of insect settlement which lies deeply buried in the history of this little group of islands.

HEMIPTERA.

CYDNIDÆ.

Pangæus bilineatus, Say.

A fore-leg only of this curious black burrowing *Cydnid* was present in the bottle of specimens. It agrees with the same organ of some individuals in my own collection; and I had previously examined a perfect specimen of this species which was brought from Bermuda by Mr. J. M. Jones.

PENTATOMIDÆ.

Nezara viridula, Linn.

One specimen, a female of the uniformly green variety, is present in the collection. This species has been widely distributed throughout most parts of the warm divisions of both the Old and the New World. It is common in North Carolina, Georgia and Florida, besides the West Indies, and it might readily have been transported to Bermuda with plants by vessels from either of the localities mentioned.

JASSIDÆ.

Cœlidia olitoria, Say.

Only the head of a specimen occurs with the other insects in the bottle. This insect is easily identifiable, but it must be regretted that the entire insect was not present, for further investigation. This little leaf-hopper is very common upon the black alder in many of the Atlantic States, and it is a matter of much interest to know upon what plant it lives in Bermuda.

HOMOPTERA.

CICADIDÆ.

Cicada tibicen, Linn.—**C. pruinora,** Say.

Said to occur on the main island. Mr. J. M. Jones says: "A very noisy individual, very appropriately named 'scissor grinder', may certainly be heard, if not seen, during the hot weather. It is a very quick sighted insect, and is difficult to capture. It remains perfectly motionless until the net is drawn towards it, when off it starts with a swift jerk and a loud buzz of derision." This is our common green *Cicada* of the United States, and it does not belong to Fidicina as the author quoted supposes.

PSEUDONEUROPTERA.

Mesothemis longipennis, Burm.

A damaged specimen is present. It proves to be a male of the strongly colored variety, with the base of the wings, especially of the posterior ones, deeply suffused with fulvous.

Lestes unguiculata Hagen.

By putting together the pieces of a disintegrated specimen, it has been possible to identify this very interesting little species. It proves to be a female, of the fully developed type of coloration, and differs in no respect from the well matured adults which are common in New Jersey and Maryland.

The foregoing are both freshwater types of the order, and must have passed through their young stages in places where suitable food could be procured. This goes to show that ponds or swamps of fresh or mildly brackish water must exist in the vicinity of the places from which these specimens were taken. Neither of them belongs to the strong-winged and widely roving Odonata, which fly without hesitation across hundreds of miles of open ocean. Possibly the progenitors of these species might have been wafted by high winds across the six hundred miles of oceanic surface between the coast of Carolina and the Bermuda Islands. We know that strong winds, blowing off the mainland of Maryland and Virginia, carry countless numbers of nearly all kinds of insects out over the ocean, and that many of these being dropped into the waves are returned to the shores by the tides and piled up in windrows along the beaches. Among these we have often found the half drowned dragon-flies mixed in with the thick piles of beetles, bugs, wasps, and flies which stretched along the line of the retreating tide.

This suggests the fact that either the tadpoles of frogs, or the larvæ of other insects, must be present in the standing water of these islands, to afford food to the voracious larvæ and nymphs of these dragon-flies.

It is extremely improbable that these are the only kinds of Odonata inhabiting the Bermudas. The swift-winged Æschnas, and some of the large and strong species of Tramea and Libellula have been seen on ships at a greater distance from the mainland than the position of these islands. We should therefore expect to find such forms as *Anax Junius*, *Tramea Carolina*, *Pantala flavescens*, and perhaps *Libellula semifasciata* and *Lepthemis hæmatogastra*, hawking over one or another part of the low districts of Bermuda, and especially in places where mosquitoes develop in greatest numbers.

DERMAPTERA.

Labidura riparia, Pallas.

Forficula gigantea, Fab.

A few specimens of the male of this large and showy earwig are present in the collection. Two of these measure 10 lines to the end of the abdomen, while the chelæ have a length of fully three lines. This species was originally derived from the region of the Mediterranean, but it has recently been widely distributed by commerce to parts along the eastern border of the United States, and no doubt the same agency has transferred it to the soil of the Bermudas.

ORTHOPTERA.

BLATTIDÆ.

Blabera Americana, Linn.

Several specimens both of the adult and larval form, are in the collection. They differ in no respect from the usual types which are now distributed over most of the world by the activities of commerce. This species is common in warehouses near the docks in some of our cities adjacent to the Atlantic coast, but it seems not to have formed a permanent lodgment in any of them. The supply is kept up by the frequent arrivals of vessels from tropical countries, in which they rest concealed between the packages of merchandise.

Panchlora Maderæ, Fab.

This is another common cockroach widely distributed by the agency of commerce. It is alluded to in the work of J. M. Jones, under the name *Blatta Maderensia*, as being seen in "cellars and other dark places, on these islands, where it is commonly known by the name of 'Knocker,' from a habit it has of making a noise like a person gently tapping a box, or skirting board."

ACRYDUDÆ.

Stenobothrus maculipennis, Scudder.

A pair of the sexes, from which the colors have been pretty thoroughly extracted by the alcohol, is present in this collection. In points of structure they correspond with specimens common to the region around Baltimore. Yet it is to be regretted that the entire absence of original color and consequent obscurity of pattern of marking make it impossible to ascertain to which one of the varieties these individuals belong.

GRYLLIDÆ.

Gryllus luctuosus, Serv.

This abundant North American cricket seems to be well settled upon the islands, although we are not informed as to its habits and distribution in that locality. It is the most littoral of our species inhabiting the Atlantic region, and finds a home in all the States from eastern Massachusetts, on Cape Cod, to the neighborhood of Saint Augustine, Florida. .

Evidence is no longer wanting as to the modifications in the length and structure of the wings and wing-covers of this species. From an examination of one colony after another on one of the beaches south of Baltimore, at intervals through a period of more than twenty years, I am led to the conclusion that the small colonies of twenty or more individuals are derived from the eggs of a single female. Several times the tide has carried off and drowned all the individuals from a short sand beach, which had to be re-stocked by another brood the succeeding year. By aiding in this work through the introduction of gravid females from other beaches, I have essentially restored the original condition of the colony. In these assemblages a small number of full-winged individuals occur almost every year, and during times in which the beach becomes clogged by excess of mud or carbonaceous matter the crickets become partly darker in color. A great advantage is

gained in studying this species, because of the great length of
its ovipositor and the greater proportions of all its organs, as
compared with its nearest relatives in the same section of
country. The black species which we have all along regarded
as *Gryllus Pennsylvanicus* Burm. lives almost within the same
territory, but it prefers the dark loamy soils farther inland,
and only ventures upon the open pale sand beaches when
hard pressed for food or moisture.

Further, the *G. luctuosus* is very variable in color and pattern of marking in the various young stages, as well as in the
fully adult. The *G. Pennsylvanicus* on the other hand is very
slightly variable in color, and has a much shorter ovipositor, but
it also has occasional individuals of both sexes fully winged.

Now the specimens brought in from Bermuda display precisely the same differences of degree and kind of color and
structure that we observe belonging to those colonies inhabiting the tidal region south and east of Baltimore. From Bermuda the long and short winged individuals are present in
both sexes, and the colors vary in both young and adults.

LOCUSTIDÆ.

Orchelimum vulgare, Harris.

A badly broken female specimen is the only representative
of this form in the collection. The length and form of the
ovipositor and the shape and markings of the head and pro-
notum definitely refer it to this species.

DIPTERA.

TABANIDÆ.

Tabanus, sp. ?

This is one of the smaller horse-flies, related to *T. lineola*, Fabr.,
but in color it resembles the *T. cincta*, Fabr. The specimens are
too much altered to be accurately determined.

Odontomyia, sp. ?

Two specimens are in the collection.

Sarcophaga carnaria, Linn.

One specimen of the usual type is present in the collection.

A CONTRIBUTION TO THE KNOWLEDGE OF THE SPIDER FAUNA OF THE BERMUDA ISLANDS.

BY

DR. GEO. MARX.

Little has been heretofore known of the spider fauna of the Bermudas. Mr. Blackwall described six species in the Ann. and Mag. of Nat. Hist., 1868; and Prof. E. Simon, in speaking of the Arachnida of the Atlantic Islands in the Annales Soc. Entom. de France, 1883, has none to add to the list of Mr. Blackwall. He, however, alludes to the character of the Bermuda spider fauna as appearing to be related to that of the Azores and the Canary Islands.

Lately, Prof. Angelo Heilprin, of Philadelphia, visited the Bermudas and collected there twelve species, and by his kindness I have been able to study this addition to the spider fauna of that region.

Mr. Blackwall described the following species:
> *Loxosceles rufescens*, Luc.
> *Epeira gracilipes*, Blackw
> *Xysticus pallidus*, Blackw
> *Salticus diversus*, Blackw.
> *Heteropoda venatoria*, Linn.
> *Filistata depressa*, Koch.

The collection of Prof. Heilprin contains the following species:
> *Uloborus Zosis*, Walck.
> *Nephila clavipes*, Koch.
> *Epeira caudata*, Hentz.
> *Epeira labyrinthea*, Hentz.
> *Theridium tepidariorum*, Koch.
> *Argyrodes nephila*, Taez.
> *Pholcus tipuloides*, Koch.
> *Dysdera crocata*, Koch.
> *Menemerus Paykullii*, Aud.
> *Menemerus melanognathus*, Luc.

Heteropoda venatoria, Linn.

Lycosa Atlantica, nov. spec.

Loxosceles rufescens, Luc., has been found in the West Indies, Central America and Florida.

Heteropoda venatoria, Linn. seems to occur, under a certain latitude, everywhere around the globe.*

Filistata depressa, synonymous (according to Simon) with *Filistata capitata*, Hentz, is quite common in the southern part of the United States.

Uloborus Zosis, Walck., is recorded from the West India Islands, Central and South America, and occurs also sometimes in southern Florida.

Nephila clavipes, Koch, is found in Brazil, Central America, Florida, Texas, and Mississippi.

Eperia caudata, Hentz, inhabits the United States from Massachusetts to Georgia.

Epeira labyrinthea, Hentz, is also common in the United States, and has been collected in the West Indies, Central and South America, as far south as the Straits of Magellan, and in California.

Theridium tepidariorum, Koch, is common to Europe and America.

Argyrodes nephila is reported from Peru, Cayenne and the Southern States of the United States.

Pholcus tipuloides, Koch, has been described by this author in his work "*Die Arachniden Australiens,*" page 281, from specimens collected at the Samoa Islands.

Dysdera crocata, Koch, is recorded from Greece, France and Germany, and is also common in the United States.

Menemerus Paykullii, Aud., and *Menemerus melanognathus*, Luc., have been found nearly everywhere on the globe.

From this material it is difficult to infer the true character of the fauna of these Islands. The frequent arrival of vessels

---

*See an article by Rev. H. C. McCook, in Procee l. Academy Nat. Sci. Philad , 1878.

from many foreign ports, the drift of the Gulf Stream, and other causes, have introduced into this region a number of species originally foreign to that locality, but which, in time, not only have acclimatized themselves, but have, in a more or less marked degree, driven away and extinguished the indigenous spider fauna.

This fact has been noticed in all localities open to the importation of a foreign element. These species are then called *cosmopolitan*, and by far the greater number of those brought to notice by Mr. Blackwall and Prof. Heilprin bear this character; but drawing a superficial conclusion from this material it seems that the spider fauna of the Bermudas is more American than anything else, for out of the seventeen species now known, only four are original (so far) to these Islands, and nine are found also in the limits of the United States.

Lycosa Atlantica, nov. spec.

Cephalothorax dark olivaceous brown, with a narrow longitudinal yellow band over the middle, which begins at the posterior margin and runs over the whole length into the region of the first eye row. Another, equally colored, but somewhat broader, band runs at the sides, above the lateral margin, terminating at the sides of the pars cephalica. Mandibles dark brown with long, thick, and black pubescence. Maxillæ and labium more reddish brown; sternum lighter, olivaceous yellow with a lighter border, hairy. Palpi and legs uniformly light olivaceous yellow, with black hairs and without rings or markings, tarsal joints of the former infuscated. Abdomen: dorsum dark olivaceous brown with a narrow, whitish, slightly spear-shaped figure, which is edged by a very narrow blackish line; behind this a row of four rather indistinct white (small) round spots, which reach the apex. Venter light yellow, middle region still lighter.

Cephalothorax as long as patella X, tibia IV; one-third longer than wide; back straight, evenly sloping in back and front, face nearly perpendicular. Lower eye row longer than second. Middle eyes of lower row about twice as large as the lateral eyes. Distance between the large eyes of the second row smaller than their diameter; eyes of the third row as large as middle ones of the first row. Mandibles as long as tibia III. Cephalothorax long, 6 mm.; broad, 4·5 in the middle region; in front, 2·2.

Abdomen, long, 6 mm. Mandibles, 2·7.

Femur I	4	Patella 2		Tibia 3·3	Metatarsus 2·8		Tarsus 2·2		Total, 14·3
"	II 3·8	"	2	" 3	"	2·7	"	2	" 13·5
"	III 3·4	"	1·8	" 2·8	"	3	"	1·5	" 12·5
"	IV 5	"	2·1	" 4	"	5·5	"	2·4	" 19

EXPLANATION OF PLATE 14.

Fig. 1.   *Uloborus Zosis*, Walck.   Female.
       1.a.   Abdomen from the side.
       1.b.   Epigynum.
       1.c.   Male palpus.
Fig. 2.   *Menemerus Paykullii*, Aud.

NOTES ON A SMALL COLLECTION OF MYRIAPODS FROM THE BERMUDA ISLANDS.

BY

CHARLES H. BOLLMAN.

The following species, which were collected by Prof. Heilprin in the Summer of 1888, although limited in number, show the diverse origin of the myriapod fauna of the Bermuda Islands. Heretofore, *Julus Moreleti* had only been found in the Azores Islands; *Mecistocephalus Guildingii* in the West Indies; *Lithobius lapidicola* in Europe; and *Spirobolus Heilprini*, by having scobina, shows its West Indian and not African origin, for all the *Spiroboli* found in the latter continent belong to the subgenus from which scobina are absent.

These four species, besides a specimen of *Scolopendra subspinipes* which I have in my collection, are all that have as yet been reported from the Bermuda Islands.

Spirobolus Heilprini, sp. nov.

Diag.—Related to *Spirobolus flavocinctus*, Karsch., but the segments very distinctly segmented; anterior part not striate; antennæ and legs reddish-brown.

Type.—Museum Acad. Nat. Sci. Phil. Greenish-black, posterior margin of segments rufous; antennæ and legs reddish-brown. Slender, anterior segments scarcely attenuated. Vertex smooth, sulcus shallow; clypeus only moderately emarginate, foveolæ 2 + 2, distant; sulcus sub-continuous with vertical. Antennæ rather slender, reaching second segment in both sexes. Ocelli arranged in a suboval or subtriangular patch, 45–55, in seven or eight series. Segments not smooth; posterior parts above with short and wavy striæ, beneath with short and straight striæ; median part with a transverse sulcus which ends above repugnatorial pore; posterior above with a few striæ, beneath almost smooth or with a few weak oblique striæ. Lateral lobes of first segment rounded, a weak marginal sulcus. Anal segment with a flat, thick mucro, which passes beyond valves; anal valves weakly margined, not punctate; anal scale obtusely angled. Repugnatorial pore placed on anterior division, small and rather deeply set. Legs extending slightly beyond sides of body. Male: slenderer than female; coxæ of 3d, 4th, and 5th pairs of legs produced into short lobes; tibia and first two tarsal joints beneath with an oval roughened lobe; joints of anterior legs short and thick, third and fourth pairs of legs strongest; tarsi without a pad; ventral plate of copulation-foot triangular, as high as foot, its base not concave, its posterior surface ridged, thus making the plate of a triangular-pyramidal form; anterior part of first foot not as high as ventral plate, triangularly pointed, the ventral plate ridge separating them; posterior part of anterior foot as high as ventral plate, its apex with a short blunt lobe on its posterior surface; posterior copulation-foot bifid, projecting out of the opening, the upper branches flattened and fan-shaped at the end, which is convex; lower branch elongate-lanceolate, its upper edge serrate; basal part of foot rectangular and white, while the upper part is yellowish.

Segments male, 46; female, 44.

Length 52^{mm}, width 3·8^{mm}- 4·2^{mm}

This species is described from six broken and badly preserved specimens. In the type of copulation-foot it resembles
S. arboreus and *S. Dugesi*, and it is very probable that all
the species belonging to this group have the same type, *i.e.* the
ventral plate triangular and as high as posterior part of anterior part, while the anterior part is less, the posterior foot
bifid and projecting out of the opening.

I have named this species after Prof. Angelo Heilprin, of the
Academy of Natural Sciences of Philadelphia.

Julus Moreleti, Lucas.

In the collection are a number of female specimens which
I refer to this species. It has only been found in the Azores
Islands.

These specimens have the striæ of the anterior division of
the segments not so irregular as is represented in Porath's figure
of the species.*

Segments 42–49. Adult almost black, legs reddish-brown;
young dusky, with a lateral row of black spots and a medium
black dorsal line, bordered with yellowish.

Mecistocephalus Guildingii, Newport.

Three specimens. These are so moulded and broken that
it is almost impossible to make much out of them; but in the
characters of the head they seem to be identical with the West
Indian species.

Lithobius lapidicola, Meinert.

Two specimens, male and female. Joints of antennæ 26;
ocelli 8 or 9, in three series; coxal pores male 2, 3, 3, 2, female
3, 4, 4, 3; spines of first pair of legs, 0, 1, 1; of penultimate
pair, 1, 3, 3, 1; of anal pair, 1, 3, 2, 0; spines of female genitalia stout, claw very distinctly tripartite, middle lobe not
much the longest; length male 7^{mm}; female 8^{mm}.

---

*Am några Myriopoder från Azorerna. Öfver. Kongl. Vet. Akad. Forh.
Stockh., 820, 1870.

It is very possible that these specimens are not identical with *L. lapidicola,* a European species; but as they are rather mutilated, I have hesitated to describe them as new.

IX.

ZOOLOGY OF THE BERMUDAS (*continued*).

MOLLUSCA.

The species of mollusks enumerated in the following pages, although probably far short of the actual number found in the region, give a good idea of the character of the molluscan fauna of the archipelago. Somewhat more than 170 marine forms, and 30 terrestrial species, are catalogued; before our visit barely more than one-half of this number had been officially recorded from the Bermudas. The general relationship of this fauna has already been discussed in Chapter V, and it is, therefore, not necessary to enter here into any further details connected with the subject.

CEPHALOPODA.

Cuttle fishes are said to be abundant in the Bermudian waters, but we were not very successful in our search after these animals. Two moderately large octopods, which we could only see, but not obtain, may possibly be the common West Indian *Octopus vulgaris*, or one of the forms that have been separated off from it as a distinct species. We made considerable efforts to capture one of these, but all our attempts to dislodge the creature from its hold upon the interior of a rock-crevice were unvailing. The following species (*Octopus chromatus*) was obtained beneath a stone on the beach of Flatts Village.

Octopus chromatus, n. sp.

Argonauta.

It seems to me likely that at least two forms of the argonaut are found here—*Argonauta hians* and *A. Argo*. Unfortunately I am compelled to rely upon my memory alone for the determination of these forms, and possibly I may be in error. The animal or parts of the animal of the argonaut have been several times captured on the Bermudian shores, and Mr. Bartram, of Stock's Point, has a beautifully preserved specimen of one of the species that was caught, I believe, some forty years ago, or more.

Spirula Peronii, Lam.

The shell of this cephalopod is very abundant, and may be found largely gathered in with the strewn Gulf-wrack which in most places lines the coast. The shells are also found in quantities, a half dozen or more, in the tidal rock-cavities, whither they had been swept by the sea.

GASTEROPODA.*

Murex erosus, Brod.

La Paz, Panama.

Murex nuceus, Mörch.

This species was originally described from the Bermudas. An identical form from Marco, west coast of the peninsula of Florida, is in the collections of the Academy of Natural Sciences of Philadelphia.

Purpura deltoidea, Lam.

Fla., Bahamas.

Purpura hæmastoma, L.

var. *P. undata*, Lam.
P. bicostalis.

Forida, W. Indies.

---

* The localities mentioned indicate in a general way the range of the species.

Sistrum nodulosum, Ads.

Florida, Aspinwall.

Triton cynocephalus. Lam.

St. Thomas, Philippines.

Triton pileare, L.

W. Indies, Philippines.

Triton chlorostoma, Lam.

W. Indies, Mauritius, Philippines.

Triton tuberosum, Lam.

W. Indies, Mauritius, Society Is.

Epidromus concinnus, Reeve.

Philippines.

This species, the specimens of which are absolutely identical with the forms from the Philippine Islands, is closely related to Tryon's *Epidromus Swifti*, from Antigua, but the ribs are less prominent than in that species, and the general outline of the shell is more acicular.

Epidromus (Triton) lanceolatus, Menke.

I give this species on the authority of Matthew Jones (Contributions to the Natural History of the Bermudas, Trans. Nova Scotia Inst.)

Ranella cruentata, Sow.

var. *R. Rhodostoma*, Beck.
      R. Thomæ, D'Orb.

St. Thomas, Mauritius, Philippines.

Fasciolaria distans, Lam.

I did not myself meet with this shell. Jones obtained a single specimen, partly imbedded in the calcareous rock, and it is therefore not unlikely that the animal is still an inhabitant of the surrounding waters.

Leucozonia cingulifera, Lam.

Florida,

Hemifusus morio, L.

Martinique, W. Africa.

? Pisania biliratum, Reeve.

I have little doubt that the species occurring under this name in Jones's list is a *Cantharus*.

Cantharus Coromandelianus, Lam.

W. Indies.

Cantharus tincta, Cour.

Florida.

Phos. sp.?

Two young specimens which I have been unable to identify with known species.

Nassa ambigua, Montf.

W. Indies.

Marginella apicina, Menke.

Florida, Bahamas.

Marginella minuta, Pfr.

Bahamas.

Volvaria avena, Valenc.

Florida, W. Indies.

? Volvaria pellucida,

Olivella oryza, Lam.

W. Indies.

Olivella nivea, Gmel.

W. Indies.

These two are probably one species.

? Olivella miliola

Oliva reticularis, Lam.

Florida, W. Indies, Venezuela.

? Oliva bullula, Sow.

The species marked with this name in Jones's list is probably *Olivella oryza* or *O. nivea.*

Dall, in his list of the West Atlantic marine Mollusca (Bull. U. S. National Museum, No. 24, 1885), includes *Oliva mutica*, Say, among his Bermudian species, and quotes Krebs as his authority. This is an error. Krebs makes no mention of the occurrence of the species in the Bermudas (Catalogue West Indian Marine Shells, p. 39).

Columbella mercatoria, L.

W. Indies.

Columbella cribraria, Lam.

Florida, Bahamas.

Columbella (Anachis) sp. ?

A form closely allied to *A. plicaria*, from New Caledonia.

Cythara (Pleurotoma), n. sp.

Identical with a form from Key Largo, Florida.

Conus Agassizii, Dall.

Strombus gigas, L.

Florida, W. Indies, S. America.

Strombus accipitrinus, Mart.

W. Indies.

All the specimens of this species that I have seen from the Bermudas lack the epidermis, and have the columellar surface of a leaden, gray color. They seem to differ from the normal type of the species in having a broader and more elevated spire, and in a more regular nodulation of the body-whorl.

Cypræa cervus, L.

W. Indies, Panama.

This species grows to a very large size. A specimen which I had the privilege of examining in the collection of Mr. Bartram, of Stock's Point, far surpasses in this respect all other specimens of the species which have come to my notice. The species occurs as a sub-fossil at St. George's and elsewhere.

Cypræa cinerea, Gmel.

Also occurs as a sub-fossil.

Trivia quadripunctata, Gray.

Bahamas.

Trivia suffusa, Gray.

Bahamas.

I believe that these two forms are merely varieties of one and the same species, being connected by a number of inter-mediate types.

Cyphoma gibbosa, L.

Florida, Cuba.

Dolium perdix, L.

W. Indies, Pacific Islands.

Natica canrena, L.

Florida, W. Indies, Costa Rica.

The nidus of probably this species, which we found in the Flatts Inlet, is a collar of lime-mud, in form very much like that of *Natica heros*, of the east American coast.

Natica lactea, Guild.

This species is very close to, if not identical with, *Natica Flamingiana*, Recl., from the Feejee Islands.

Natica Marochinensis, Gmel.

W. Indies, W. Africa, Sandwich Islands.

Crepidula convexa, Say.

Eastern United States.

Two small specimens obtained through the kindness of Miss A. Peniston, which are seemingly this species.

! Crepidula fornicata, L. (young.)

Phorus agglutinans, L.

This is probably a rare shell. We obtained but a single specimen from the rocks off Bailey's Bay.

Adeorbis.

A number of small specimens, obtained from the sands of the north shore, which closely resemble the *Helix* (*Adeorbis*) *cyclostomoides* of Pfeiffer, from Cuba. The spire in the Bermudian form is, however, practically wanting.

Scalaria clathrus, L.

S. lamellosa, Lam.
S. coronata.
South Carolina, W. Indies, Europe.

Ianthina communis, Lam.

Florida.
Common on both shores of the islands, and largely associated with *Spirula Peronii*. Its occurrence away from the immediate shore-line is, doubtless, due to wind deposit.

Ianthina globosa, Swn.

Florida.
This species is apparently much less abundant than the preceding.

Turritella, sp.?

Several fragments of a species which I have been unable to determine.

Petaloconchus, sp.?

! Vermetus Knorri, Say.

? Vermetus radicula, Stimpson.

Siliquaria (Tenagodus) ruber, Schum.

I give this species as a member of the Bermudian fauna on the authority of Mörch.

Cæcum Floridanum, Stimps.

Florida.
A single specimen from the sands of the north shore.

Cæcum termes, n. sp.

A number of specimens obtained through the kindness of Miss Peniston,

† Eulima Jamaicensis, Adams.

W. Indies.

Leiostraca. sp.?

Stylifer, sp.?

Abundant in the skin of the large holothurians (*Stichopus*). I failed to obtain specimens, and am, in consequence, ignorant of the species.

Turbonilla pulohella, D'Orb.

Chemnitzia pulchella, D'Orb.
St. Thomas.

Turbonilla pusilla, Adams.

St. Thomas.

Littorina zio-zao, Chemn.

Bahamas, Cuba.

Littorina angulifera, Lam.

Littorina lineata, Gmel.
et. L. scabra, L. ?
W. Indies.

Littorina murioata, L.

W. Indies.

Littorina Mauritiana, Lam.

Tectarius Antillarum, D'Orb.

W. Indies.

Modulus Floridanus, Conr.

Florida.
The *Modulus lenticularis* (Chemn.) of Jones is probably this species.

Modulus pisum, Beck.

Placed here on the authority of Morch.

Litiopa striata, Rang.

Cerithium literatum, Brug.

W. Indies.

Cerithium ferrugineum, Say.

C. versicolor, Adams.
C. Bermudæ, Sowb.
C. Eriense, Val.
Florida, W. Indies.

Cerithium, sp.?

A form apparently closely related to *Cerithium diadema,* Watson, from Madeira.

Potamides minimus, Gmel.

P. nigrescens, Menke.
P. albovittatus, Adams.
?*P. zonale* (in part), Brug.
E. United States, W. Indies, Europe.

Triforis turris-Thomæ, Chemn.

St. Thomas.

Triforis intermedius, Adams.

St. Thomas.

Rissoina crassicostata, Adams.

St. Thomas.

Rissoina Sagraiana, D'Orb.

St. Thomas.
Rissoina micans, Adams, in Jones's list

Hydrobia?

Nerita peloronta, L.

W. Indies.

Nerita tessellata, Gmel.

W. Indies.

Nerita versicolor, Gmel.

Florida, W. Indies, Costa Rica.

Neritina viridis, L.

W. Indies, Lancerote, Mediterranean.

Neritina Virginea, L.

W. Indies, Brazil.

Phasianella Kochi, Phil.

South Africa.

This species is not unlikely but a variety of *Phasianella pulla*, L., from the Mediterranean, the west coast of Africa, etc.

Turbo pica, L.

W. Indies, Nicaragua, Panama.

This large and beautiful shell is abundant along the southern shores, where it also occurs imbedded as a sub-fossil in the calcareous rock. It appears to be always largely water-worn, and in no instance did it contain the animal. Indeed, I am assured by old residents who have intentionally sought for the animal that it has never been found in the island-waters. Probably the animal inhabits the deeper waters, and only the empty shell is cast upon the beach.

Astralium longispina, Lam.

Trochus (Imperator) calcar, L., of Jones's list.

W. Indies, Honduras.

Stomatia picta, D'Orb.

St. Thomas.

Fissurella viridula, Lam.

W. Indies.

Fissurella Barbadensis, Gmel.

Barbados.

Fissurella Antillarum, D'Orb.

Cuba.

Fissurella alternata, Say.

Eastern United States, Nicaragua.

Emarginula dentigera, n. sp.

Emarginula pileum, n. sp.

Patella notata, Lam.

Patella confusa.
W. Indies.

Bulla striata, Brug.

Florida, W. Indies, Africa, etc.

Bullina (Bulla) nitidula, Dillw.

Hydatina (Bulla) physis, L.

These two West Indian species are here enumerated on the authority of Jones.

Utriculus Candei, D'Orb.

Florida, W. Indies.

Umbrella (Operculatum) Bermudensis, Mörch.

On the identification of Mörch.

Siphonaria picta, Sow.

Panama, Mazatlan.

? Siphonaria alternata, Say.

Florida.

Dentalium semistriatum, Guild.

W. Indies.

Cadulus quadridentatus, Dall.

Florida coast.

Chiton squamosus, L.

Chiton marmoratus, Gmel.

Ischnochiton purpurascens, Adams.

Tonicia Schrammi, Shuttl.

Aplysia æquorea, n. sp.

Chromodoris zebra, n. sp.

LAMELLIBRANCHIATA.

Teredo,   sp. ?

Found in driftwood; species undetermined.

Martesia striata, L.

On the authority of Jones.

Siliqua bidentata, Spengl.

Florida, Panama.

Macha Antillarum, D'Orb.

Lyonsia Beaui, D'Orb.

Florida, W. Indies.

Semele variegata, Lam.

Florida.

Semele orbiculata, Sow.

Florida, S. Car., W. Indies.
? *Semele reticulata*, Say.

Semele cancellata. Sow.

St. Thomas.

Asaphis coccinea, Martyn.

W. Indies.

Strigilla flexuosa, Say.

W. Indies.

Tellina polita. Say.

East coast America, Florida.

Tellina Gouldii, Hanley.

California.

Tellina Tampæensis, Conr., from Florida, is probably identical with this species; but there is no question as to the identity of the Bermudian and Californian forms.

Tellina Gruneri, Phil.

Tellina intastriata, Say.
Capsa spectabilis of Jones's list.
St. Thomas.

Tellina radiata, L.

and var. *Tellina unimaculata*.
W. Indies.

Tellina interrupta, Wood.

Tellina maculosa, Lam.

St. Thomas.

Tellina lævigata, L.

Bahamas.

Tellina magna, Spengl.

Florida, W. Indies.

Tellina exilis, Lam.

Guadeloupe, Jamaica.

Macoma eborea, n. sp.

Chione Venetiana, Lam.

St. Thomas, France.

Circe Cubaniana, D'Orb.

Florida, W. Indies.

Venus cancellata, L.

Given on the authority of Jones.

Cytherea Penistoni, n. sp.

Cardium serratum, Brug.

Florida, St. Thomas.

Chama macrophylla, Chemn.

W. Indies.

Chama exogyra, Conr.

California, Mexico.

Two specimens, which are absolutely undistinguishable from the species of the west American coast.

Chama lingua-felis, Reeve.

Chama Bermudensis, n. sp.

Lucina edentula, L.

Loripes chrysostoma of Jones's list.

Florida, W. Indies, Nicaragua.

Lucina squamosa, Lam.

? *Lucina imbricatula*, Adams.
W. Indies.

Lucina tigrina, L.

W. Indies.

Lucina imbricatula, Adams.

W. Indies, Aspinwall.

Lucina Antillarum, Reeve.

Lucina costata, D'Orb.
W. Indies.

Lucina obliqua, Reeve.

Lucina pectinata, Ads.
St. Thomas.

Mysia pellucida, n. sp.

A form identical with the above, from St. Thomas, is contained in the collections of the Academy of Natural Sciences.

Crassatella Guadelupensis, D'Orb.

W. Indies.

Arca Noæ, L.

W. Indies, Europe.

Arca Deshayesii, Hanley.

St. Thomas.

I have no doubt that the *Arca Americana*, of Gray, mentioned by Jones as a very common shell of the Bermudas, is this species.

Arca gradata, Brod.

Arca Domingensis, Lam.
Florida, St. Thomas, Mazatlan, Cape Verde I.

Arca Adamsi, Shuttl.

Florida, Cuba.

The Bermudian specimens appear to be somewhat less produced posteriorly than the more southern forms, but in other respects they agree very closely. The species is probably identical with *Arca solida*, B. and S., from the Californian coast.

Arca imbricata, Brug.

W. Indies, Mazatlan, Feejee Islands.

The *Arca mutabilis*, of Sowerby, does not differ essentially from this species.

Mytilus exustus, L.

Mytilus Domingensis of Jones's list.

Modiola tulipa, L.

E. United States, Bahamas, W. Indies.

Lithodomus Antillarum, Phil.

Lithodomus niger.

Florida, St. Thomas.

Lithodomus appendiculatus, Phil.

Cuba.

Crenella lateralis, Say.

S. Carolina, Florida.

Avicula ala-perdicis, Reeve.

Meleagrina placunoides, Reeve, of Jones's list.

Florida, W. Indies.

Avicula Atlantica, Lam.

S. Carolina.

? *Avicula macroptera*, Lam.—W. Indies, E. Indies.

Perna ephippium, L.

Antilles, Indian and Pacific Oceans.

Pinna rudis, Lam.

W. Indies.

Plicatula ramosa, L.

Florida, W. Indies.

Lima tenera, Chemn.

Lima fragilis, Sow., of Jones's list.

Florida.

Pecten zig-zag, L.

W. Indies, Pacific Ocean.

Pecten imbricatus, Gmel.

St. Thomas, Pacific Ocean, Australia.

Spondylus Americanus, Lam.

W. Indies.

There seems to be much confusion regarding the species of Spondylus which inhabit the West Indian and Bermudian waters, and, as it appears to me, an unnecessary multiplication of specific names. The individual variation among the species is very great, and does not permit of that close characterization of forms which has been attempted by some systematists. I believe that *Spondylus Americanus* includes much, if not all, of what has been described under *Spondylus longitudinalis, S. ustulatus, S. coccineus,* and *S. crinaceus.*

Anomia ephippium, L.

America, Europe.

Ostrea frons, L.

W. Indies.

PULMONATA.

The land shells enumerated in the following list are in excess by some ten forms of the number hitherto credited to the Bermudas. Most of these we collected ourselves or obtained through the aid of local collectors; others I found in the collection made a few years ago by Mr. G. Brown Goode, while a few have been quite recently sent to me by Miss A. Peniston, of Peniston's, Bermuda. The remaining forms are given on the authority of Fischer (*Manuel de Conchyliologie*), Bland (quoted by Wallace in "Island Life," pp. 256–57), and Pfeiffer. These are preceded by an asterisk. For the determination of the species I am largely indebted to Mr. H. A. Pilsbry, Conservator of the Conchological Department of the Academy of Natural Sciences, who has also prepared a special paper on the more

distinctive Bermudian Helices. The localities placed opposite
the species indicate in a general way their geographical distribution.

Helix (Cochlicella) ventricosa, Drap.

Mediterranean region, Canaries, Azores.

Helix (Microphysa) vortex, Pfr.

W. Indies, S. United States.

Helix (Polygyra) microdonta, Desh.

Texas, Florida?

Helix (Hyalosagda) discrepans, Pfr.

* **Helix (Vallonia) pulchella, Müll.**

Europe, Azores.

Given on the authority of Bland, as quoted by Wallace. This
species appears to differ but little, if at all, from *Helix minuta*,
of Say, a common form of the United States.

Helix appressa, Say.

Pennsylvania to Illinois and Arkansas.

! **Helix hypolepta, Shuttl.**

This species, which appears to have been obtained from the
Bermudian collection of M. Bland, is very inadequately described by Shuttleworth (*Diagnosen neuer Mollusken*, p. 129), so
that it is barely recognizable as a distinct species. It is closely
related to *Helix minuscula*, of Binney.

Pœcilozonites Bermudensis, Pfr.

Pœcilozonites circumfirmatus, Redf.

Pœcilozonites Reinianus, Pfr.

Succinea Bermudensis, Pfr.

* **Succinea fulgens, Lea.**

Cuba.

* **Succinea margarita, Pfr.**

Hayti.

Pupa Jamaicensis, Adams.

Jamaica.

Pupa pellucida.

W. Indies, Yucatan.

* Pupa Barbadensis, Pfr.

Barbados.

On the authority of Bland, quoted by Wallace.

Pupoides fallax, Say.

? *Bulimulus nitidulus,* Pfr.

United States, W. Indies.

Bulimulus decollatus, L.

Cuba, Southern Europe, etc.

This species appears to have been recently introduced, and is not mentioned in the earlier lists of the Bermudian pulmonates. It is, however, very abundant in places; we found it especially numerous along the roadside near the Smith Parish Church.

* Cæcilianella (Cionella) acicula, Müll.

Central Europe, Madeira.

Stenogyra octona, Chemn.

Antilles, Caracas, Panama.

Alexia Bermudensis, Ads.

Pedipes tridens, Pfr.

This species is considered by Arango (*Contribucion a la Fauna Malacologica Cubana,* p. 60) to be identical with *Pedipes mirabilis,* Mühlf., a form from Cuba, Jamaica, Guadeloupe, etc.

Melampus pusillus, Gmel.

W. Indies.

Melampus Redfieldi, Pfr.

Melampus coffea, L.

W. Indies, Mexico, Guiana.

Melampus (Tralia) cingulatus, Pfr.

Cuba, Jamaica, Porto Rico.

Truncatella Caribæensis, Sow.

Cuba, Jamaica.

Truncatella subcylindrica, Gray.

Cuba, St. Thomas, Porto Rico.

? Truncatella pulchella, Pfr.

Cuba, St. Thomas, Porto Rico.

Helicina convexa, Pfr.

Onchidium trans-Atlanticum, n. sp.

NEW SPECIES OF MOLLUSCA.

Octopus chromatus. (Pl. 15, fig. 1.)

Body spheroidal, somewhat acuminate behind, and impressed, but not furrowed, ventrally; mantle opening extending about one-half around the circumference of the body, and terminating some distance below and back of the eyes. The head not much narrower than the body; eyes not conspicuous, with a wart above each; funnel largely free, reaching about half way to the base of the web, which is about as long as the body and head combined.

Arms longest as 1. 3. 2. 4, although possibly the second pair outmeasured the third pair previous to contraction; slender, very tapering, and exceedingly attenuated toward the apex suckers fairly large, closely placed, and in regular zigzag alternation from the base, contracting with a quadrangular outline.

Body granulated posteriorly, and to a less extent in the region of the neck. Color milky, closely blotched or speckled with ochre, giving a yellowish appearance, and sprinkled with brown.

Length of specimen about nine or ten inches.

The only form with which I can closely compare this species is the *Octopus Bermudensis* of Hoyle (Challenger Reports, Zoology, XVI, p. 94, Pl. II, fig. 5), which is described from a single young specimen, measuring, including the arms, not more than two and a-half or three inches. It differs from this form in the extremely tapering and attenuated arms, their relative lengths (1. 3. 2. 4 instead 1. 2. 3. 4), and in the disposition of the acetabula, which are in zigzig alternation from first almost to last; the body is also in part granulated, and the siphon, instead of being attached for nearly its full length, is largely free.

I should have hesitated, perhaps, in describing this as a new species, distinct from *O. Bermudensis*, and preferred supposing that the characters indicated by Hoyle were not very clearly marked, or that they possibly represented only the immature form, but Hoyle distinctly states that while his specimen is probably immature, the characters are so well marked as to safely permit of their recognition as typical of a new species (*op. cit.*, p. 95).

Aplysia æquorea. (Pl. 15, figs. 2, 2a, 2b).

Body broadly oval, with a moderately elongated neck; tentacles cylindrical, slit at the extremity; buccal lobes broad, infolded; mouth between fairly developed lips; aperture to opercular cavity on a slightly raised papilla.

Color drab or greenish; exterior surface with thin black annulations and irregular markings, which are few and scattered; the inside of the mantle-lobes, as well as the cover to the opercular cavity, almost free of blotches.

Shell narrowly-elongate, somewhat oblique, and calcareously lined; longitudinally radiated, and transversely finely striated.

Length of animal about four and a-half inches.

A single specimen, found in shallow water on the south side of Castle Harbor, opposite Tucker's Town.

The nearest ally of this species is probably the *Aplysia ocellata* of D'Orbigny, from the Canary Islands, or the common *A. dactylomela*, from the eastern Atlantic, of which the former is by some authors considered to be only a local variety (Rochebrune, *Nouvelles Archives du Muséum*, 1881, p. 264). From both of these forms, apart from other characters, it differs in the absence of the heavy ocellation, and from *A. dactylomela* in lacking the purple lining on the mantle margins. From *A. ocellata,* again, it is clearly marked off by the non-maculated surface of the interior of the mantle lobes and of the opercular covering. The shell in the Bermudian form is comparatively narrower than in any other large species of *Aplysia* with which I am acquainted, and wholly different in outline from that of either of the two species above referred to. I have fully satisfied myself on this point through an examination not only of the figures furnished by Rang and D'Orbigny but of actual specimens.

Dobson, in a communication made before the Linnæan Society of London (Jour. Linn. Soc., Zoology, xv, p. 159, *et seq.*, 1881), identifies a specimen of *Aplysia* from the Bermudas with the *A. dactylomela*, and describes the color as being "a rich drab, marked all over with circles and streaks of velvet black, the latter most abundant on the mantle covering the shell and on the lateral swimming lobes. The shell agrees in all respects with that of *A. dactylomela* as figured by Rang, and the only difference observable is that the margins of the swimming lobes are not tinged with violet. This might be accounted for by supposing that such a fugitive color had disappeared in the alcohol, but the captor does not remember to have seen it in the living animal." This may be the true *A. dactylomela* or *A. ocellata*, but it is, doubtless, distinct from the species above described. I am confirmed in this supposition by the examination of a specimen recently collected by Prof. Dolley in the Bahamas, and which has been placed in my hands through the kindness of Prof. Leidy. This Bahaman form has the massive ocellation and blotching distinctive of *A. ocellata* or *A. dactylomela*, and further agrees with these two species (or varieties) in the form of the shell. The stellate opening to the opercular cavity appears to be destitute of a papilla. This is the form, probably, that Mr. Dobson received through Surgeon R. Vacy Ash.

Deshayes described some years ago an Aplysia, ocellated and of a yellowish color, from Guadeloupe (*Journal de Conchyliologie*, 2d. ser., ii, p. 140) under the name of *Aplysia Schrammii*, but the species is so imperfectly characterized that it is almost impossible to determine its exact relationships.

Chromodoris zebra. (Pl. 15, figs. 3, 3a.)

Animal of the form typical of the genus; head portion considerably extended and expanded in motion; caudal portion moderately elongated; base flattened; mantle beaded immediately over the tail.

Color bright blue above, variously lined and streaked with light yellow; on the dorsal surface the yellow markings are disposed in longitudinal wavy or nearly straight lines, one or more specially prominent lines along the dorso-lateral border. Sides of animal irregularly reticulated or angulated with yellow markings; under surface pale blue, bordered with faint yellow.

Rhinophores deep indigo or black, the rhinophoral aperture bordered with yellow; gills 12 or 13, black, bordered with yellow, and carrying blue cilia; under surface of head blue, with yellow spots.

Length, when expanded, three and a-half inches.

Three specimens, dredged in about ten fathoms on the north side of Harrington Sound. I dissected one of these and found that the stomach is lodged entirely within the mass of the liver. The alimentary canal is sharply deflected forward (dorsally) beyond the buccal or œsophageal tracts, and is caught up in a nerve ring proceeding from the supra-œsophageal ganglia.

This species appears to be third or fourth of the genus found in the western Atlantic. It differs clearly from the *C. picturata* of Mörch (*C. Mörchii*, Bergh, *Mus. Godef.*, part xiv) and *C. gonatophora* of Bergh, two West Indian species. In the scheme of coloring the species appears to be nearest to *Doris pulcherrima* of Cantraine (*Malacologie Méditerranéenne*, p. 57, Pl. 3, fig. 6, = *D. Villafranca?* of Risso), from which, however, it differs in a number of details, such as the number of gills, etc.

Onchidium (Onchidiella) trans-Atlanticum. (Pl. 15, figs. 4, 4a.)

Body convex, smoke color or dark olive; lighter, dirty or greyish-green on the under surface; pedal disk considerably more than one-third the width of base, yellowish-green; mouth margin papillose, bunchy; under surface obscurely or obsoletely tuberculose; dorsal surface closely verrucose, with finer granules interspersed between the warts.

Anal aperture immediately beyond the extremity of foot, infra-marginal to a raised border; respiratory orifice between the anal pore and the apex of body.

Length about three-quarters of an inch.

About a dozen specimens, found in a rock hollow on the north shore just beyond Wistowe near Flatts Village, at an elevation of about two feet above the water.

This is, as far as I am aware, the only species of *Onchidium* that has thus far been recorded from the western Atlantic. Its occurrence is, therefore, of considerable interest as bearing upon the subject of geographical distribution. Nearly all the species of the genus are confined to the Eurafrican and Indo-Pacific waters, although one species is known from Arctic America, one from the California coast, and one from the west coast of South America (Bergh, in Semper's *Reisen im Archipel d. Philippinen*, Land Mollusks, VI).

The Bermudian species appears to be most nearly related to *O. Carpenteri*, from the California coast, but differs from it in color. The positions of the anal and respiratory apertures differ from what is indicated by Stearns (Proc. Acad. Nat. Sci. Phila., 1878) to exist in the west American form, although agreeing with the determinations made by Bergh for manifestly the same species.

Emarginula dentigera. Pl. 17, fig. 7.

Shell flattened, scutiform, broadened posteriorly, and with the apex slightly sub-central; surface with radiating ribs, which alternate in size—sometimes two smaller ones between each pair of larger ribs—and project (the larger ones) prominently beyond the general margin. The impressed concentric lines give to the ribs a knobbed appearance. Fissure fairly long, narrow.

Length nearly .25 inch.; height .1 inch.

Emarginula pileum. Pl. 17, fig. 6.

Shell elevated, with the form of a Phrygian cap; apex largely posterior, well beaked; radiating lines alternate, deeply impressed by the concentric lines of growth, and appearing knobbed. Fissure moderately long, parallel-sided, and occupying the position of one of the larger ribs.

Length slightly exceeding a quarter of an inch; height .2 inch.

Cæcum termes. Pl. 17, fig. 5.

Shell arcuate, gradually increasing in size anteriorly, where it is somewhat swollen; surface longitudinally costated, the costæ appearing slightly rugose near the swollen base through the passing of the lines of growth. Mucro distinct, well excentric.

Color of shell yellowish; surface glossy.

Length, about .1 inch.

This form appears to be fairly abundant. It may be readily distinguished from most of the other longitudinal costated species by the very nearly equal diameter of the shell, which is only slightly swollen basally.

Chama Bermudensis. Pl. 17, fig. 1.

Shell thick, ponderous, sub-cordiform; the right valve considerably smaller than the left, but yet thicker and deeper than in most Chamas; beak of left valve prominent, spirally incurved; ligamental furrow in left valve deep, arciform; dental sulcus broad, moderately deep, and supported inferiorly by a prominent plate tooth.

Muscular impressions elongated, not deep. External surface roughly corrugated by the lines of growth.

Height of shell (left valve), measured to the top of beak, about three inches; length, measured along the antero-posterior axis, 2.3 inches; thickness of single valve 1.2–1.5 inch.

Dredged in large quantities in Harrington Sound.

Macoma eborea. Pl. 17, fig. 2.

Shell moderately inequilateral, truncated in the lower half;

posterior flattened. Right valve with widely diverging cardinal teeth, the space between which receives the double-tooth (grooved medially) of the left valve; lateral teeth in right valve.

Pallial sinus large, extending more than half across the shell; external surface (white) concentrically and delicately lined by the lines of growth.

Length of shell three-quarters of an inch; height, two-thirds of an inch.

Mysia pellucida. Pl. 17, fig. 3.

Shell thin, convex, ovally orbicular; the umbones moderately prominent; no lunule; hinge-line linear, a single medially-grooved cardinal tooth in the left valve (resembling Felania).

Adductor impressions oval, not much elongated. Shell white, nearly smooth.

Length of single specimen somewhat over a half-inch; height the same.

Cytherea Penistoni, Pl. 17, fig. 4.

Shell sub-trigonal, the beaks prominently elevated; lunule cordiform; the dental characters normally those of the genus; anterior lateral tooth (left valve) triangular, lamellar.

Margin of shell even; pallial sinus broad, directed upwards, and not quite reaching the centre of the shell. Lines of growth closely set, and even. Shell covered with a chestnut epidermis; interior purplish.

Length, .6 inch; height, .5 inch.

It gives me pleasure to name this delicate, and abundantly represented, Cytherea after my friend Miss. A. Peniston, of Peniston's, from whom I have received much valuable assistance in the preparation of my material illustrating the Bermudian fauna.

ON THE HELICOID LAND MOLLUSKS OF BERMUDA.

BY

H. A. PILSBRY.

Through the courtesy of Professor Angelo Heilprin I have been enabled to study the Bermudian land shells collected by the party conducted by him during the past summer.   Among them were examples of all the helicoid species which have been reported by previous observers from the island, some containing the living animal.   The species, with the exception of a number of artificially introduced European shells, are mostly forms well known from various West Indian localities, such as *Helix cereolus* var. *microdonta* Desh., *H. vortex* Pfr. and others; but besides these, there are a number of shells peculiar to Bermuda, and these last have furnished material for the following notes.

The helicoid species confined to Bermuda are as follows: *H. Bermudensis* Pfr., *H. Nelsoni* Bld., *H. Reiniana* Pfr., *H. circumfirmata* Redf., *H. discrepans* Pfr.   As to the systematic position of these forms there has been considerable difference of opinion among authors; the first, *H. Bermudensis*, has been placed in *Caracolus* by Von Martens, in *Hyalina* by Clessin, in *Zonites* by Bland; *H. Reiniana* has been considered a *Patula* by Pfeiffer, Clessin, Tryon and Fischer; and *H. circumfirmata* and *discrepans* have been placed in *Microphysa* by Von Martens and Binney, in *Hyalosagda* by Clessin, Tryon and others.

Thus it will be seen that these species have been distributed into several genera in two distinct families.   Upon examining the soft parts, however, I find that all have essentially the same organization and without doubt belong to the same genus.

Dr. O. Boettger proposed in 1884, for the lower Miocene fossil *Helix imbricata* Braun, and the *H. Bermudensis* Pfr., the name of *Pœcilozonites*.   He gave no diagnosis of the new group, but assigned it a position between the typical Palæarctic

Zonites and the American groups *Zonyalina* and *Moreletia*, a position which the anatomical characters prove to be erroneous.[1]

By error, the genus was quoted "Poecilozonites *Sandberger*" in the Zoological Record for 1884, and this error was repeated by Tryon[2], who gives the first diagnosis of the group published, with *H. Bermudensis* as the type and only species. We may then consider the *H. Bermudensis* Pfr. the type species of the genus. Whether the *H. imbricata* Braun be associated with the Bermudian shells or not is a point still to be settled. The superficial resemblance is marked; but, as the history of the species of *Poecilozonites* teaches us, "systematizing" helicoid land mollusks by the shells alone is the merest guess-work.

The fact that the fossil species which Dr. Boettger proposes to unite with the Bermudian form is from the Lower Miocene formation of Germany, is in itself no great objection to the view that they are congeneric; for no fact is better established

---

[1] " * * * Endlich sei noch einer nahen Verwandten der Hocheimer untermiocäner *Helix imbricata* Al. Braun gedacht, die Sandberger bekanntlich zu *Trochomorpha* (*Discus*) gestellt hat. Ich gebe die Aehnlichkeit zu; aber zur Section Videna H. u A. Adams, Discus Alb., möchte ich die betreffende fossile Art nur ungern stellen, da alle mir bekannten lebenden Arten dieser Gruppe zum mindester einer verdichter basalrand, der oft recht erheblich Helix-artig umgeschlagen ist (wie z. B bei *Tr. Merziana* Pfr.) besitzen. Viel näher liegt daher wohl der vergleich der *Helix imbricata* mit der etwas kleineren, mit zwei braunen Bändern gezierten *Hyalinia Bermudensis* Pfr. von den Bermudas, deren Uebereinstimmung in allen wesentlichen Charakteren bei directem Vergleich sofort in die Augen springen dürfte. Freilich kommen wir hier fast von dem Regen in die Traufe, da die systematische Stellung die-er lebenden Art selbst noch in hohen Grad unsicher ist, was ihr Autor durch ein vorgesetzes '?' sehr richtig selbst schon angedeutet hat. Bei *Hyalinia* kann sie unmöglich bleiben. Da sie meiner Ansicht nach auch nicht in die indische, indomalayische und polynesische gattung *Trochomorpha* passt, so dürfte eine eigene Gruppe fur *Hyal. Bermudensis* und *Helix imbricata* zu errichten sein, für welche ich den Namen *Poecilozonites* vorschlage, und die ich am liebsten zwischen die ächten paläarktischen *Zonites* und die tropisch-amerikanischen Gruppen *Moreletia* und *Zonyalina* vorläufig als Section in der Gattung *Zonites* Montf. einreihen möchte, bis die Anatomie der lebenden Art eine mehr gesicherte Stellung in System an die Hand geben wird." O. Boettger in *Neues Jahrbuch fur Mineralogie, Geologie u. Palaeontologie*, 1884, ii Bd., s. 139.

[2] Manual of Conchology, 2d. series, iii, p. 19, 95.

in malaco-geography than the close affinity existing between the European Tertiary land Mollusca and those now inhabiting the West Indies.[1] To explain this relationship existing between two regions separated by the whole expanse of the Atlantic various theories have been offered. One of the most plausible is that which bridges the Atlantic by an ancient (Eocene, early and Middle Miocene) continent—an *Atlantis*. This view has been advocated by the well-known conchologist Dr. W. Kobelt[2] and by others.

But although this theory explains many anomalies in the distribution of mollusks, I must freely confess that the objections to it seem to me almost insurmountable. The recent work of the Challenger, Blake, and other deep-sea explorations, all tend to confirm the view held by Guyot, Dana, Agassiz and others, that the great oceanic basins, practically as they exist to-day, are of great antiquity, and render the existence of a former Atlantic continent with any considerable Western extension, highly improbable.

A view more in accordance with the facts with which we are at present acquainted seems to me to be the following: It is a well ascertained truth that until towards the close of the Miocene, large portions of Northern Africa as well as Europe were submerged; and it appears probable that the westward flowing equatorial current of the Indian Ocean extended across northern Africa, and united with the Atlantic northern equatorial current, which now flows westward from northern Africa through the Antilles into the Gulf of Mexico. This current would afford a means of transport not only for the free swimming embryos of marine mollusks (and there are not

---

[1] This affinity, although doubtless very great, has been considerably exaggerated. There is, for instance, no warrant for referring European Tertiary species to the exclusively New World genera *Pleurocera*, *Anculosa*, *Tulotoma*, *Mesodon*, *Carinifex*, *Melantho*, and others. There seems to have been no infusion of European Tertiary types into the North American snail fauna east of the Californian region. This fauna is truly autochthonous.

[2] *Nachrichtsblatt d. deutschen Malak. Gesell.*, 1887, p. 147

a few forms both of gasteropods and pelecypods common to the Mediterranean and Gulf Provinces), but also, through the agency of floating materials, trees, etc., swept from rivers, land mollusks may have been transported across the Atlantic, just as they have been carried by the Gulf Stream from the West Indies to the outlying island of Bermuda,[1] a distance of over 700 miles.

A further development of the same idea explains certain peculiarties in the distribution of species common to the Pacific and the Gulf of Mexico. The presence of Miocene and Pliocene deposits render it certain that there was communication between the Gulf and the Pacific across the Isthmus of Panama as late as the Pliocene. And a portion of the equatorial current probably swept directly through to the Pacific. Thus it is likely that those forms common to both sides of the isthmus, will prove to be of Atlantic origin, and to have been distributed westward.

The indigenous Bermudian mollusk-fauna, marine as well as terrrestrial, has undoubtedly been derived wholly[2] from the West Indies. And since the island is typically oceanic, "a solitary peak rising abruptly from a base only 120 miles in diameter," surrounded on all sides by between 2500 and 3000 fathoms depth, we have an idication here that land mollusks of many families, *Helicidæ*, *Zonitidæ*, *Succinidæ*, *Pupidæ*, *Helicinidæ*, even *Vaginulidæ* (for a large undescribed species of *Vaginulus* exists upon the island), may be transported far out to sea, and, in all probability, by the agencies mentioned above.

The considerable divergence existing between the various species of the zonitoid genus peculiar to Bermuda, *Poecilozonites*, indicates that the island is of considerable antiquity. We may define the genus as follows:

---

[1] See Darwin, Origin of Species, 6th ed., p. 353. Also a paper by Mr. C. T. Simpson, On the Distribution of Land and Fresh-water Shells in the Tropics, Conch. Ex. ii, p. 37, 50.

[2] See on this point the chapter on the "Relationship of the Bermudian Fauna," *ant.*, p. 88. A. H.

PŒCILOZONITES.

Generic characters: Shell helicoid, subtrochiform, depressed conic, or subdiscoidal, perforate or umbilicate, obliquely striate, ornamented with radiating zigzag flammules or spiral bands of chestnut color on a lighter ground; whorls numerous (7–10), very slowly widening; body-whorl more or less flattened or compressed below the usually carinate periphery, not descending anteriorly; aperture more or less irregularly lunate; peristome simple, the columellar margin slightly expanded and thickened with a white callus which encircles the pillar within. Animal similar in form to *Helix*; foot narrow, short posteriorly, scarcely reaching behind the shell, and without longitudinal furrows above its margin or caudal mucous pore; orifice of genitalia on the right side of neck, near to, but not under, the mantle; mantle margin simple; jaw like that of *Limax*, very thin, arcuate, with a broad blunt median projection anteriorly; radula with tricuspid central teeth having quadrate basal plates, the central cusps projecting beyond the anterior margins of the basal plates, the side cusps rather short, with well reflexed cutting points; lateral teeth similar but asymmetrical, lacking the inner cusps; marginal teeth aculeate, with simple thorn-shaped cusps and oval basal plates.

It will be seen by the above definition that the genus cannot be included in any of the groups with which its species have been associated by authors; the zonitoid dentition at once removes it from the *Helicidæ*, and the absence of a caudal mucous pore, the more anterior position of the orifice of the genitalia and the coloration of the shell, separate it from *Zonites* and its subgenera.

The relationship of the species of *Pœcilozonites* to one another is shown by the similarity of the radulæ and jaws, and of the external characters of the animal; in the shells, which at first glance seem to be a heterogenous assemblage, by the callus which coats the columella, the compression of the whorl below the periphery, and especially by the color-pattern, which is the same in all the species, consisting of zigzag flammules radiating from the sutures. In *P. Bermudensis* the flammules coalesce into continuous bands above and below the periphery in the adult; but an examination of young specimens reveals the same pattern that is found in *P. circumfirmata, P. Reiniana*, etc. The internal spiral lamella of *P. circumfirmata* would incline one at first to separate it from the other species; but it is scarcely of generic importance, in view of the fact that in all other characters the species is very similar to *P. Bermudensis*, etc.

The following analysis shows the inter-relations of the various species:

A.   Base of shell with a revolving lamina within.

circumfirmatus, discrepans.

B.   Base of shell without lamina.

    a. Aperture rounded below; umbilicus wide   *Reinianus.*

    b. Aperture angulate below; umbilicus narrow

Bermudensis, Nelsoni.

Pœcilozonites Bermudensis, Pfr.   (Pl. 16, figs. E. c.)

The typical species is a form of about twenty-five mm. diameter, solid, coarsely irregularly striate and acutely carinate at the periphery; a broad chestnut band usually encircles the shell above the periphery, and another below it, but these are sometimes absent; the inner whorls of the spire usually retain traces of the original color-pattern of radiating flames, and the base in young examples is radiately streaked (Pl. 16, fig. E). The base is convex, and not indented around the narrow and deep umbilicus, but is angulated at its margin; the parietal wall is generally covered by a shining white layer, with which the interior of the shell is lined. Reeve, Tryon and other authors have figured the shell of this species.

The jaw is like that of *P. circumfirmatus.*

The radula (Pl. 16, fig. c) is rather long. The central teeth have basal plates almost as broad as long, the median cusps projecting below their lower margins, with well-developed cutting points; the side cusps short, attaining about the middle of the basal plate, and directed outward; the lateral teeth are similar, but lack inner cusps; they are about eight in number, and are followed by about four transition teeth; the marginals number about fifty on either side, their cusps become more slender toward the outer edge, and the basal plates shorter. A central tooth, with five adjacent lateral teeth, and a group of transition teeth, with a true marginal tooth, are shown in the figure.

Helix albella of Chemnitz (not of Linnæus) and *H. ochroleuca* of Pfeiffer (not Ferussac) are, I believe, synonymous with this species. The former is placed in *Eurycratera* in Pfeiffer's *Nomenclator*, and the latter has been compared to *Pachystyla rufozonata*, a form somewhat similar in characters of the shell, but belonging, of course, to a distinct group.

Pœcilozonites Nelsoni, Bland. (Pl. 16, figs. J, K, L.)

A fossil form, differing from *Bermudensis* in its much greater size, the greater number of whorls, more convex base, coarser striation, impressed sutures, and especially in the peculiarly prominent dome-shaped upper whorls. These are, indeed, so closely coiled as to resemble a specimen of *P. circumfirmatus*. The coloration, imperfectly shown in several specimens before me, is that of *Bermudensis;* and whilst its affinities are with the latter species, I regard it as a divergent branch, rather than as an ancestor of that form.

As has been observed in other cases of species approaching extinction, and probably subject to some decided and unfavorable changes of environment (in this case, perhaps, due to the comparatively recent subsidence and partial submergence of the island*), the shell exhibits great mutations and distortions of form; sometimes the spire is elevated conical, sometimes much depressed; frequently the planes of the upper and lower volutions are not parallel, and the spire is consequently canted

---

* See Challenger Report, Narrative, vol. i, p. 138.

to one side.   The species is remarkably large, solid and roughly
sculptured for a zonitoid.

Pœcilozonites Reinianus, Pfr.   (Pl. 16, fig. 1.)

This heretofore unfigured species is discoidal in form, widely
umbilicate; the umbilicus about one-third the diameter of the
base, and exhibiting all the whorls; the apical whorl is smooth
and whitish; the following whorls are quite convex, with deep
sutures, brownish, very prettily zigzagly flammulate with
chestnut color, like many of the species of *Patula.*   The body-
whorl in adult examples is rounded; the base concave around
the umbilicus, and the general aspect that of *Patula.*

The jaw is like that of *P. circumfirmatus.*

The radula (Pl. 16, fig. D) is similar to that of *P. Bermudensis*
except in the following points: the cusps are larger, with much
more widely reflexed cutting points; the perfect lateral teeth
are seven on either side; the change to marginals is quite
abrupt, as there are but two real transition teeth; the marginals number about sixteen on each side, the inner six or
seven of about equal size, the outer ones rapidly decreasing
toward the edge.   The basal plates are longer than in the
other species.   A central tooth with two adjacent laterals and
one marginal are shown in the figure.

Pœcilozonites Reinianus Pfr. var. **Goodei** Pilsbry.

This form is similar in coloration and texture to *P. Reinianus.*
It is more broadly umbilicated, planorboid, the spire flat, or
even sub-immersed; whorls six.

Alt. 3, diam. 10 mill.

Among the Bermudian shells sent to Prof. Heilprin from the
U. S. Nat. Mus. were a number of this variety, which seems to
me distinct enough for a name.   The types of the variety are
No. 94,424 of the National Museum register, collected by G.
Brown Goode.

Pœcilozonites Bermudensis Pfr.

The result of my dissection of this species was a surprise to
me, for I had expected the same form of genitalia as is found

in *Zonites*. The penis is rather short, *convoluted*, thick, the vas deferens inserted at its termination, is rather short. The cloaca is large, wide ; below the penis there is a long club-shaped sac, its base dilated where it enters the cloaca. This is proably a dart-sack, although the specimens examined by me contained no dart. On the penis near its base arises a duct, which uniting with another arising opposite the penis, is continued into a long duct coiled around the vagina, and ends in a small oval bulb, the receptaculum seminis or spermatheca. The albumen gland, etc., offer no unusual characters. I did not dissect out the ovo-testis. My specimens were quite hard, having been in strong spirit.

The connection of the duct of the spermatheca with the penis is unique, so far as I know, in the Pulmonata, and suggests the probability of self-impregnation.

Mr. W. G. Binney has kindly called my attention to his note upon the dentition and jaw of *II. Bermudensis* and the dentition of *II. circumfirmata* in the Ann. N. Y. Acad. Sci., iii, p. 86, 105. The first species is placed by him with doubt in *Zonites* with the remark that "it seems to belong to no described genus." *II. circumfirmata* is left in *Microphysa*, for want of a better place, but Mr. Binney points out the fact that the species belongs to the *Vitrinea* rather than to the *Helicea*.

Pœcilozonites circumfirmatus Redfield (Pl. 16, fig. F).

A form with much the appearance of *Hyalosagda*, a group with which it has been classed by some authors. It is a delicate, subtranslucent, yellowish-brown shell, marked with brown streaks, spots and flammules; the whorls are separated by moderately impressed sutures; the apex is like that of *P. Reinianus ;* the last whorl is more or less angulate around the periphery, rather flattened below the angle, then convex, indented around the narrow, deeply perforating umbilicus; there is a white calcareous deposit around the columella, inside, as in the other species, and an acute white lamella which revolves within the base near to the periphery, a character which none of the preceding species possess. The variation in form is very great—specimens more elevated than my figure F being not infrequent, and these are connected by examples more and more depressed (fig. G) with the flattened lenticular form called by Pfeiffer *H. discrepans*. This extremely depressed variety, now figured for the first time (Pl. 16, fig. H.), cannot be considered specifically distinct from the *P. circumfirmatus*.

Jaw (Pl. 16, fig. B) transparent, very thin, arcuate, with blunt extremities and a wide obtuse median projection below.

Radula (Pl. 16, fig. A) as described for *P. Bermudensis*, but with only seven laterals, two or three transition teeth, and about twenty-eight marginals. The marginals have longer basal plates than in *P. Bermudensis*.

Helix (Microphysa) hypolepta Shuttleworth.

Of this minute form no diagnoses or figures have been published, although the name has been upon the lists for many years. The shell was apparently unknown to Pfeiffer except by the remarks of Shuttleworth, who says under his diagnosis of *H. minuscula* Binn.: "Altera species proxima, sed testa aperte umbilicata, et anfr. ultimo basi devio distincta, in insula Bermuda occurrit, cujus specimina plurima ab am. Bland accepi, atque *H. hypolepta* nominavi."

The shell is minute, discoidal, whitish, subtranslucent and shining, with wrinkles of increment above, nearly smooth beneath. The four whorls are very convex, quite gradually widening, the last one with the periphery above its middle, the lower lateral surfaces sloping somewhat as in *H. vortex* Pfr. The aperture is small, not very oblique, oval. The lip is acute, upper and basal margins quite arcuate, the baso-columellar margin slightly expanded. The umbilicus is broad, more than one-third the diameter of the shell.

Alt. 1, diam. 2¼ mill.

It is evidently allied to *H. (Microphysa) vortex* Pfr., but is much smaller, flatter, with broader umbilicus. I need not compare *Zonites minusculus* with this shell; a glance at the figures will show at once the difference.

Helix hypolepta. Shuttleworth, Diagnosen neuer Mollusken, no. 6, from the Bern. Mittheil., March, 1854, p. 129.

The group *Microphysa*, in which I have placed this shell, has been a stumbling block to most of the authors who have recognized it. It consists of small, umbilicated, thin, hyaline shells, with sharp lip to the lunar-oval aperture, convex whorls and impressed sutures. There is little in all this to separate it from certain forms of *Zonites* (*Z. minusculus*, for example). But the *Zonites* have narrow aculeate marginal teeth to the radula, while these shells, typified by *H. Boothiana* Pfr., have the dentition of *Patula*. The marginal teeth are low and wide, with several denticles.

APPENDIX.

NOTES ON THE RECENT LITERATURE OF CORAL REEFS.

W. J. L. Wharton. "Coral Formations." Nature, Feb. 23, 1888.

The author cites a number of submerged atoll-like banks in the China Sea, depressed to depths of 30–60 fathoms, on the rims of which the corals are still in active growth. Of such are the Tizard Bank (with a length of 32 nautical miles, and a depth of water over the rim of 4–10 fathoms, and in the centre of 30–47 fathoms), the Prince Consort Shoal, and the great Macclesfield Bank, the last, 70 miles in length, and covered, in its deepest part, with sixty fathoms of water. These are given as evidences of banks that are being built up through coral-growth, and which are ultimately supposed to reach the surface. But the author gives no evidence to show that these banks are not in reality banks of subsidence, drowned atolls, similar to what Mr. Darwin considered the Chagos Banks to be. The fact that corals are still growing on the rim in the one case and not in the other, does not affect the question.

Capt. Wharton disputes Mr. Murray's conclusion that the great depth of atoll lagoons can be formed through simple aqueous solution, and observes: "but the fact that for large areas it [the surface of the reef] remains awash, and must have so remained for ages, seems to me to point to the supposition that the removal of matter is too insignificant to account for the formation of deep lagoon channels in this manner, though doubtless it may explain the shallow pools and creeks found in all fringing reefs."

J. Murray. "Coral Formations." Nature, Mch. 1, 1888.

A purely theoretical answer to the objections contained in

the paper of Capt. Wharton, noticed above, relative to the formation of deep lagoons through solution. No facts bearing on the subject are given.

G. C. Bourne. "Coral Formations." Nature, Mch. 1, 1888.

The author coincides with the views of Capt. Wharton as to the inefficacy of solution in producing deep lagoons. "It has seemed to me, as it has to him, that the solution of dead coral rock in the interior of a reef does not sufficiently account for the formation of lagoons, and that the true cause of the atoll and barrier lagoons surrounded either by a reef which is awash, or by a strip of low land, lies in the peculiarly favorable conditions for coral growth present on the steep external slopes of the reef." The favorable conditions are supposed to be due to the action of currents on coral growths [*not* a better food-supply], currents of moderate strength " not so strong as to dash them [the corals] to pieces, but strong enough to prevent deposition of sand. Such conditions are found everywhere on the external slopes," where the "main part of the current flows tangentially around the obstruction," adds to "greatest advantage around the periphery of a reef," and forms a ring-shaped reef; "no theory of solution is required to explain the central depression."

Mr. Bourne, as a non-believer in the theories of solution and subsidence, fails, however, to explain how the ring-form was constructed below the zone of coral-growth ; the extension of the lagoon far below this line remains unaccounted for.

R. Irvine. "Coral Formations." Nature, Mch. 15, 1888.

An attempt to determine the rate of solution of lime-carbonate in the sea. From experiments made on the genus Porites (coral), using sea-water with a specific gravity of 1.0265, and a temperature of from 70° F. to 80° F., the author arrives at the conclusion that "dead or rotten coral exposed to sea-water under these circumstances is soluble to the extent of 5 to 20 ounces per ton." For further computation the author assumes a reef with a lagoon already formed, half a mile in diameter. "This will give an area of about 600,000 square yards, and supposing the water to be 3 feet deep and only one-sixth part of this to be in actual contact with the dead coral, we have 100,000 tons exerting its solvent action. This would give, were the sixth part of the lagoon water to be expelled and replaced with fresh sea-water at each tide, and taking the solvent action at only 10 ounces to each ton, an amount of carbonate of lime removed equal to about 3000 tons each year."

Mr. Irvine curiously asserts that while he does "not insist that such an amount of carbonate of lime *must* year by year be removed from the lagoon," he yet thinks that the "experiments show that the carbonate of lime so removed may easily exceed any additions to the lagoon by secretions of animals living in it, or by coral sand carried into it by wind and waves from the outer edge in the same space of time, and therefore I think the balance of evidence is in favor of Mr. Murray's explanation of lagoon formation."

But Mr. Irvine does not inform us on what grounds he assumes that this internal waste may exceed accumulation or accretion by growth. The argument is of that nature which assumes that a "large" figure can accomplish anything, or cover a multitude of omissions. The removal of 3000 tons of material annually from a comparatively small basin appears like a large amount, but when this quantity is closely scrutinized its vastness largely disappears. A ton of limestone, allowing a weight of 150 pounds to a cubic foot, is the equivalent in a general way of 15 cubic feet; 3000 tons will therefore represent 45,000 cubic feet of material. This amount distributed over an area of 5,400,000 square feet (600,000 yards, as assumed by Mr. Irvine) would cover it to a depth of the $\frac{1}{120}$ of a foot, or the $\frac{1}{10}$ of an inch. In other words, this $\frac{1}{10}$ inch represents the annual waste according to Irvine; it is the equivalents of a cubical block of rock of 36 feet dimensions. Whether this amount is sufficient to satisfy the demands of the "solutionists" or not, I am not in a position to say; but from my observations of the waste of the Bermuda-lagoon shores, and the organic accumulation taking place over the floor of the lagoon, I am positive that it does not by a long way meet the case of these islands. The height of the shores in the Bermudas, doubtless, permits of vastly excessive destruction, and the conditions, possibly, cannot be absolutely compared with what we find in other coral islands. Nevertheless, I am inclined to believe that the organic accumulation (sea-urchins, shells, corals, Foraminifera) alone over the floor of the lagoons fully covers the quantity demanded in the computation. Bourne and Wharton are likewise of the opinion that the amount of accumulation is in excess of that of solution; the observations of these investigators were made on low-land reefs, in which the special conditions of the Bermudas were wanting.

J. G. **Ross.** "Coral Formations." Nature, Mch. 15, 1888.

Also an attempt to determine the solubility of calcium carbonate in sea-water. Mr. Ross finds that a specimen of *Oculina varicosa*, one of the hard West Indian corals, measuring about 8 square inches of surface (with a weight of 16·3164 grammes) lost by solution in 20 days 0·0748 gramme; and, similarly, a specimen of the porous *Madrepora scabrosa*, from the Feejee Islands, with a surface of 16 square inches, and a weight of 21·8540 grammes, lost in a period of 30 days 0·1497 gramme. From a circular lagoon, four miles in diameter (or with a superficial area of some 12½ miles), it is concluded that at this rate of solution there would be removed within a year 8472 tons. This if evenly distributed "would give a thickness of half an inch covering the whole area of the lagoon."

T. Mellard Reade has pointed out the error in this calculation (Nature, Apl. 5, 1888), which assumes for the quantity of lime-carbonate removed by solution 125 times that which is carried by the proposition. In other words, the removal of 8472 tons from the floor of the lagoon in question would only increase its depth (per annum) by the $\frac{1}{250}$ inch instead of one-half inch. At this rate, as Mr. Reade points out, it would require a period of a million years to hollow out a lagoon of 60 fathoms depth. There are probably few geologists who will permit such a long period for the formation of this accessory structure in a coral-reef. And if the lagoon itself is so ancient, how old must be the structure in which it is implanted? I have discussed this subject on pp. 57–59, and have, I believed, demonstrated that according to the determinations of the quantity of organic and inorganic lime-sediment contained in sea-water it would require a period of 100,000 years to build up the thickness of a single foot from the oceanic abyss. In shallow water, on the contrary, the process of construction may be a very rapid one.

H. B. Guppy. "Coral Formations." Nature, Mch. 15, 1888.

The author defines the conditions governing the form and the life of reefs as follows: "On the outer side of a reef we have the directing influence of the currents, the increased food-supply, the action of the breakers, etc. In the interior of a reef we have the repressive influence of sand and sediment, the boring of the numerous organisms that find a home on each coral block, the solvent agency of the carbonic-acid in the sea-water, and the tidal scour. These are all real agencies, and we only differ as to the relative importance we attach to each." No new facts bearing on these points are given.

G. C. Bourne. "The Atoll of Diego Garcia and the Coral Formations of the Indian Ocean." Nature, April 5, 1888.

A description of the southernmost atoll of the Chagos Group, with considerations bearing on the structure of the other reefs and coral islands of the Indian Ocean. The main facts contained in this paper, as well as those contained in the more elaborate article published by the same author in the Proc. Royal Soc., XLIII, 1888, are discussed in the body of this work. Mr. Bourne finds evidence of an elevation of some 4 feet in the Diego Garcia reef, and hence concludes that the fact precludes "the idea of any subsidence being in progress, as Mr. Darwin fancied to be the case in the Keeling atoll." The raised atolls—"atolls whose dry land just rises above the waves and submerged banks"—of the coral formations north

of Madagascar are considered to be "proof that atolls are
formed in areas of elevation, and if the facts which I have
already stated concerning Diego Garcia are of any weight, it
would seem that most of the coral formations of the Indian
Ocean mark areas of elevation rather than of rest, certainly
they are not evidence of subsidence." That the last movement in the region may have been one of elevation need not
be disputed; and, as far as any general theory of coral formation is concerned, the same movement, or a reversed one, may
be taking place to-day. But Mr. Bourne does *not* show that
the characteristic structure of the islands under special consideration was not formed during a period of subsidence, or
that no subsidence has really taken place; the fact that elevation may be now taking place in no way precludes the possibility of an antecedent subsidence. The raised marine strata
of continental areas might as well be taken in evidence of non-
submergence or subsidence. It would indeed be difficult to
prove, from what evidence Mr. Bourne has placed before us,
that Diego Garcia is not to-day subsiding, instead of rising,
despite the positive proof that is given of a recent elevation of
four feet. Assuming the correctness of Mr. Darwin's hypothesis of subsidence I fail to see what condition would be brought
about by a change of movement—*i.e.*, if such subsidence as
caused the formation of "drowned-atolls" were followed by ele-
tion—other than that which is presented by Diego Garcia and
the other reefs which Mr. Bourne describes.

Mr. Bourne does not believe that the solution-theory of the
formation of lagoons is tenable, and he challenges "the statement that the destructive agencies within an atoll or a submerged bank are in excess of the construction" (*vid. ant.*, note).

R. Irvine. "Coral Formations." Nature, Apl. 26, 1888.

The author furnishes the following results as to the solubility
of different coral fragments (and other limestones) in sea-water,
in grammes per litre, for an exposure of 12 hours: dead
Porites, 0·395; coral sand 0·032; Bermuda harbor-mud 0·041;
Isophyllia dipsacca, from Bermuda, 0·041; *Millepora ramosa*

(Bermuda) 0·036; *Madrepora aspera* 0·073; *Porites clavaria* (Bermuda) 0·093; weathered oyster-shells, 0·331; crystallized carbonate of lime, 0·123; amorphous carbonate of lime, 0·649. The rate of solution here given is vastly in excess of the results obtained by Mr. Ross.

J. Murray. "On the Structure and Origin of Coral Reefs and Islands." Proc. Royal Soc. Edinburgh, X, 1880.

An exposition of the non-subsidence or accretion theory of the formation of coral structures. The author thus sums up his conclusions (p. 517):

1. Foundations have been prepared for barrier reefs and atolls by the disintegration of volcanic islands, and by the building up of submarine volcanoes by the deposition on their summits of organic and other sediments.

2. The chief food of the coral consists of the abundant pelagic life of the tropical regions, and the extensive solvent action of sea-water is shown by the removal of the carbonate of lime shells of these surface organisms from all the greater depths of the ocean.

3. When coral plantations build up from submarine banks they assume an atoll form, owing to the more abundant supply of food to the outer margins, and the removal of dead coral from the interior portions by currents and by the action of the carbonic acid dissolved in sea-water.

4. Barrier reefs have built out from the shore on a foundation of volcanic debris or on a talus of coral blocks, coral sediment, and pelagic shells, and the lagoon channel is formed in the same way as a lagoon.

5. It is not necessary to call in subsidence to explain any of the characteristic features of barrier reefs or atolls, and all these features would exist alike in areas of slow elevation, of rest, or of slow subsidence.

The above constitute the main propositions of what is frequently termed the "Murray theory" of the formation of coral structures. These have already been discussed in the chapter on "The Coral-Reef Problem," and therefore call for no special consideration in this place. Mr. Murray, like most of the authorities who reject the Darwinian hypothesis of subsidence, gives no satisfactory data in support of his propositions (*e.g.*, 1, 3), and he appears to be satisfied with the mere possibility (doubtless to him, probability) of the correctness of the substitute theory. Nor are any facts given to indicate that subsidence has not taken place, although it is apparently considered more convenient to "do away with the great and general subsidences required by Darwin's theory." But why? In what respect is a long-continued subsidence more difficult to be believed in than an equally long continued elevation? Yet Mr. Bourne, one of the strong upholders of the non-subsidence theory, affirms his belief (*vid. ant.*) that "atolls are formed in areas of elevation" and that "most of the coral formations of the Indian Ocean mark areas of elevation rather than of rest"! Is the question then reduced to one simply of elevation or subsidence?

Mr. Murray informs us that his views "are in harmony with Dana's views of the great antiquity and permanence of the great ocean basin, which all recent deep-sea researches appear to support." It is a little difficult to see just how they are in harmony with these views, and apparently they are much less so than is the subsidence theory. Dana himself states (Report Wilkes Exploring Expedition; A. J. Science, 3d ser., XXX, pp. 94, 97) that the course of the coral islands in the Pacific conforms largely "with the axial line of greatest depression," and that the deep-water area or trough which extends southeastward from Japan through the Central Pacific conforms "well to the suggestion of the Darwinian theory." I fail to see how, if the coral growths are planted either on ascending or stable areas, the condition specially agrees with any theory of oceanic permanency. I should rather think the reverse, for permanency in the ocean would seemingly be established through progressive subsidence. Mr. Murray, however, states that all the volcanic regions which we know have in the main been areas of elevation and we would expect the same to hold good in those vast and permanent hollows of the earth which are occupied by the waters of the ocean (*loc. cit.*, p. 516). But in what lies the evidence for these assertions? It would probably be as difficult to prove a general elevation in volcanic tracts as it would be difficult to furnish that evidence in favor of subsidence in coral areas which the opponents to the Darwinian hypothesis demand. Indeed, it is well known that by many geologists volcanic tracts are considered to be areas of subsidence, rather than the reverse. This is the view now held by the foremost Austrian geologists, like Suess and Neumayr, who associate the great crustal fractures or depressions —the "sunken basins"—with volcanic phenomena. While subsidence may, and with little doubt does, initiate volcanic outflows, it seems reasonable to suppose that any very great extravasation of material from the earth's interior will produce subsidence, except in so far as this subsidence may be locally balanced by the material ejected. Dana, indeed informs us, from a study of the deep-sea soundings of the "Tuscarora" and "Challenger," that the region of the great island of Hawaii, "although it is now actively volcanic and has little growing coral about it," has seemingly "undergone more subsidence than the coral reef end of the chain, and that its height and steepness of submarine slopes are due to the fact that its outflows of lava have kept ahead of the subsidence, and also built up nearly 14,000 feet above the sea" (A. J. Science, 3*d* ser., XXX, p. 101).

H. B. Guppy. "Notes on the Characters and Mode of Formation of the Coral-Reefs of the Solomon Islands." Proc. Royal Soc. Edinburgh, 1885–86, pp. 857–904.

This is one of the most comprehensive papers dealing with the coral formations of any one single group of islands. The region of the Solomon Islands comprises, according to this investigator, all three forms of reefs—atolls, fringing-reefs, and barrier-reefs—and thus presents special advantages for the study of the coral-reef problem. The author's main conclusions may be briefly summed up as follows:

1. Reefs appear at the surface as the result of growth at about the sea-level or through upheaval.

2. The numerous detached submerged coral-shoals, which represent the early condition of reef-structures, are not able to raise themselves to within the constructive power of the breakers without the aid of a movement of elevation. Being arrested in their upward growth, at depths varying between 5 and 10 fathoms, according to the exposed or protected character of their situation, they form flat shoals of no great size.

3. Atolls of small size, *i. e.*, a mile or so across, do not assume their characteristic form until they have reached the surface. A small flat-topped shoal is first brought by upheaval to or above the sea-level; lateral extensions or wings grow out on either side, so as to ultimately form a horse-shoe reef. Such a reef presents its convexity against the prevailing surface-currents, to which in truth it owes its shape.

4. The larger atolls have probably assumed their form beneath the surface, "since, according to the principle laid down by Mr. Murray, they would then have a relatively smaller periphery for the supply of food and sediment to the interior than would be possessed by the small submerged shoals above described."

5. The true "growing edge" of a reef is the seaward slope which extends outward between the depths of 4–5 and 12–18 fathoms; where this submarine slope is more than 10° or 12°, "as is usually the case," the sand and gravel arising from oceanic degradation—which, with a more gentle slope, accumulates at its base—is carried far beyond the depths in which reef-corals thrive. In the case of reefs possessing a gradual seaward slope, *i. e.*, less than 5°, the lower margin of this band of detritus will lie within the zone of reef-building corals, and in consequence a line of barrier-reef will be ultimately formed beyond this band with a deep water channel inside. Successive series or belts of barrier-reefs thus formed may be brought to the surface through a progressive rise of the sea-bottom.

6. Reef-building corals are not restricted to a superficial

zone of 100–120 feet; under favorable conditions "they may thrive in depths of 50 or 60 fathoms, and thus we can readily explain the apparently abnormal depths inside some atolls and barrier-reefs."

7. Reefs grow out on their own talus.

It will thus be seen that Mr. Guppy dissents from those who hold to the theory of subsidence, but it can scarcely be said that his facts are fully, or even largely, in accord with the substitute theory of Mr. Murray; nor can they be said to be opposed to the requirements of the Darwinian hypothesis. Perhaps the most important of Mr. Guppy's generalizations is that reef-building corals can thrive at considerably greater depths than has been generally supposed, reaching under favorable conditions to fully three times the depth of the commonly accepted limit. Indeed, if this condition can be proved to exist it would naturally do away with much of the necessity for a belief in subsidence, since it would (or could) explain one of the most distinctive features of coral structures, the deep lagoons and channels. But the evidence on this point is of a very unsatisfactory nature. Sporadic growths of reef-building corals may well be found in depths exceeding the so-called coral-zone, but until it can be shown that anything like a reef-development takes place in this greater depth, we are justified in restricting the coral-zone to the narrow limits which have been generally assumed by naturalists. Mr. Guppy, indeed, informs us that "under favorable conditions, reef-corals may thrive in depths of 50 or 60 fathoms" (p. 903), but this statement seems to rest merely upon an antecedent statement (p. 887) that "off the reef of Choiseul Bay I [the author] did not seem to have reached this lower limit [of coral growth] in soundings of 40 fathoms." And does this indicated depth of 40 fathoms rest— as it certainly seems to—on the fact that in a cast of 31 fathoms the arming preserved a "rounded impression of the size of a billiard-ball, the inner surface of which retained the prints of small cells as if of a Porites" (Ann. Mag. Nat. Hist., June, 1884, p. 464)?

Between thriving at depths of 50–60 fathoms and the finding of an obscure impression at 31 fathoms there is surely a vast difference. But Mr. Guppy himself informs us (" Coral Soundings in the Solomon Islands," Ann. Mag. Nat. Hist., June, 1884), that in Selwyn Bay, on the west side of Ugi Island, the depth at which coral thrives is between 20 and 25 fathoms (p. 461); in Port Mary, Santa Anna, the limit is placed at 20–30 fathoms, although the deepest recognizable impression (of an Astræan) was obtained from only 17 fathoms (p. 461); off Onua the " lower limit at which coral thrives " is about 20 fathoms (p. 463); while off the northwest coast of Balãlai Island, Bougainville Straits, " a depth of 15 fathoms apparently represented the lowest limit of the zone of-corals " (p. 463).

Mr. Guppy's own observations are, therefore, practically confirmatory of the observations of nearly all other investigators who had preceded him. In fact, if we except the impression obtained at 31 fathoms, they are seemingly absolutely confirmatory; moreover, the impression may have been that of a dead coral.

Dana well remarks (" Corals and Coral Islands," 1872, p. 118) that " soundings with reference to this subject are liable to be incorrectly reported by persons who have not particularly studied living zoophytes. It is of the utmost importance, in order that an observation supposed to prove the occurrence of living coral should be of any value, that fragments should be brought up for examination, in order that it may be unequivocally determined whether the corals are living or not. Dead corals may make impressions on a lead as perfectly as living ones."

It is on this slender basis, if it is a basis at all, that Mr. Guppy constructs his theory for the formation of barrier-reefs (which inclose deep channels) and his explanation of the deep lagoons of atolls. Prof. Bonney has, it appears to me, well answered that " till Mr. Guppy can produce cases of *growing reefs* at depths well exceeding 25 fathoms, isolated instances of the occurrence, at such depths, of living corals which are among the reef-builders do not really help him; and that till he can do this he is only supporting hypothesis by hypothesis" (Nature, July 4, 1889).* The same objection probably holds to any inference being drawn from the discovery of a number of reef-genera (Stylophora, Astræa, Pavonia, Cycloseris, Leptoseris, Stephanaria, Psammocora, Montipora, Alveopora, Rhodaræa in depths exceeding 30 fathoms off the Tizard and Macclesfield banks, as reported by Bassett-Smith (Nature, July 4, 1889). At all events, more detailed information than we now possess regarding this seemingly important find is needed before satisfactory conclusions can be based upon it.

There is another point with reference to the existence of what might be called the second or deeper zone of coral growth, which is supposed to be separated by a barren sand area from the normal zone (100–120 feet), upon which Mr. Guppy is not very clear. It is assumed that the sand resulting from oceanic degradation destroys the life over which it is largely precipitated, and that were it not for its bad influence corals would be found growing on the deeper sea-ward slope as they are found growing above. The first part of the proposition is in a measure doubtless true, but the second does not necessarily follow; on the contrary, the fact that these corals are practically never found in the "barren" area is almost positive evidence against the truth of the proposition. Otherwise we should find scattered and luxuriant growths just as we find them in the interiors of the sand-swept lagoons. According to Guppy the lagoons and lagoon channels of the Solomon Islands are largely occupied "by sand and chalky mud; but in the shallower portions, and especially in those situations which are near the breaks in the reef, corals thrive in great profusion" (Proc. Royal Soc. Edinburgh, 1885–86, p. 861).

In the lagoon of Oima the individual coral-colonies are described as being very much larger than they are on the outer slope of the reef; "large masses of Porites ranged from 10 to 16

---

* I regret that up to the time of printing it has been impossible for me to secure a copy of the new edition of Darwin's work on coral islands, edited by Prof. Bonney.

feet in diameter; whilst the largest masses that I found in the wash of the breakers at the outer edge of the reef, which belonged to species of Cœloria and Mæandrina, measured only 5 feet across" (p. 890). Surely, with such evidence before us it cannot reasonably be supposed that there could be such an extermination from the outer slope, if reef-building corals really thrive at these depths, as Mr. Guppy would lead us to suppose.

With regard to the formation of the deep lagoons and channels and the actual thickness of the coral-made rock, Mr. Guppy's own views seem to be in conflict. The author apparently inclines to the views of Murray and his followers that these deep bodies of water are after-formations, and that they have been produced through steady removals of material. As factors in this removal he cites the action of carbonated waters ("solution theory") and the various forms of organic degradation (pp. 893–97).

But no instance is cited where any considerable depth of water has been brought about in this way; it is merely the assumed hypothesis of possibility. On the other hand, we are positively informed (pp. 878–79) that the lagoon of the Oima atoll (which measures nearly two miles in its longest diameter), with a depth of some 20 fathoms, is filling up through the accumulation of sand! And this condition exists in an atoll which has seemingly experienced no "upheaval since the commencement of its growth."

The same condition prevails in the case of the Keeling atoll, where, as Mr. Guppy informs us, "the lagoon is rapidly filling up with sand and coral" (Nature, Jan. 3, 1889). The facts are thus clearly opposed to the theory that is assumed.

One of the points that have been specially insisted upon by the opponents of the subsidence theory as being destructive of that theory is the supposed thinness of the coral-made rock, and much stress has been laid upon the researches of Guppy in the Solomon Islands. This subject has been considered in the chapter dealing with the "Coral-Reef Problem," but a few ad-

ditional remarks are here necessary. In his more recent pub-
lication on the "Solomon Islands" Mr. Guppy informs us that
the thickness of the coral limestone in the upraised reefs is in a
general way between 100 and 150 feet, and that he never found
an island "that exhibited a greater thickness of coral-limestone than 150 feet or at the very outside 200 feet" (p. 71)
This is in itself an important observation, but it is just what
we should expect to find in a region of elevation, as we are informed this one is. Without subsidence I fail to see how, on
the Darwinian hypothesis, a coral limestone could have a
greater thickness than 100–150 feet. The special significance
of the observation lies only in the fact that the same thickness
of coral-rock is associated with what is assumed to be a raised
atoll—namely, the island of Santa Anna. This island is described as being nearly circular in form, with a length and
breadth of two and a half and two miles respectively, and consisting "of a central basin surrounded by an elevated rim [100
to 200 feet in height], which is wanting at the middle of the
west or lee side. The bottom of the basin, which extends
downward to about 100 feet below the sea-level, is occupied
by two fresh-water lakes," the largest of which measures about
half a mile in length, and has a depth of 18 fathoms in its
deepest portion. The highest elevation of the island, a volcanic peak, 470 feet in height, rises from the rim of the eastern
border, while another elevation, of 160 feet, is found in the
centre of the depressed basin.

It does not appear clear that this is a true atoll; and Mr.
Guppy himself admits that the island differs "from the typical
reef of this description," although agreeing with the atoll-like
structures of the Solomon group ("Solomon Islands," p. 113).
It is manifestly a part of that class of structures, the horse-shoe
shaped reefs, which "do not assume their characteristic form
until they have reached the surface," and which the author
broadly distinguishes from the large atolls, which have probably "assumed their form beneath the surface" (Proc. Royal Soc.
Edinburgh, 1885–86, p. 900). "A small flat-topped shoal is

first brought up by upheaval to or above the sea-level. Lateral extensions or wings grow out on either side, so as to ultimately form a horse-shoe reef. Such a reef presents its convexity against the prevailing surface currents, to which in truth it owes its shape " (*loc. cit.*, p. 900 ; this view of the formation of atollons or horse-shoe reefs is further elaborated in Mr. Guppy's paper " Preliminary Note on Keeling Atoll," Nature, Jan. 3, 1889). Such are seemingly the conditions that we find on Santa Anna Island, but the examination of the 100-fathom contour line, which closely conforms to the actual boundaries of the island, even to the indentation of the 17-fathom Port Mary—concerning which Mr. Guppy expresses himself as having " been unable to obtain any satisfactory explanation " (" Solomon Islands," p. 117)—proves conclusively, I believe, that the surface exposed above water is merely the correspondent of that which is below it, in other words, the island has grown up on a base of its own form, which base is seemingly a breached crateral cone of a volcano. It repeats on a larger scale what is still presented by its own highest elevation, the eastern volcanic cone, which carries " a small circular hollow, between 100 and 150 yards across and 35 or 40 feet in depth. There was a time in its history, when the present summit alone appeared at the surface of the sea as a tiny ring of coral reef, capping a submerged volcanic peak, the remains of which still exist in the shallow basin on the highest part of the island " (*op. cit.*, p. 113). I think we are well justified from this evidence in assuming that the large breached-ring is similarly only an upgrowth from a larger crateral border, upon which the small cone is perched. Mr. Darwin early recognized the possibility of such a structure, and he guardedly affirmed his belief that under suitable conditions a " reef like a perfectly characterized atoll " might be formed over the rim of a crater (" Structure and Distribution of Coral-Reefs," 1842, p. 89).

It is therefore in no way surprising that the thickness of the coral-made rock on this island should be comparatively slight, and nowhere exceeding 150 feet.

Mr. Guppy ingeniously argues from the character of the rock which in many of the islands immediately underlies the coral-limestone, and which in certain organic and mineral features recalls the deeper deposits of the ocean, that the amount of elevation in the region has been very great, and that the coral formations are planted directly upon deep-sea or even abysmal deposits. Thus, it is claimed in the history of Santa Anna Island that "a submerged volcanic peak, lying at a depth of probably 2000 fathoms below the surface, was covered by a deep-sea mud, and then elevated until it became the base of a coral atoll, which has been subsequently upheaved together with its foundations to a height of nearly 500 feet above the sea" ("Solomon Islands," p. 113). I fail, however, to see the force of the argument. In the first place, it is well known that the pelagic organisms which contribute their remains to the deep-sea deposits are largely—if not, indeed almost wholly—animal forms which inhabit the superficial zone of the sea; likewise, the inorganic substances which accumulate at the bottom—cosmic dust, disintegrated pumice, etc.—are derived from the upper regions. Hence, manfestly, a shallow open-sea deposit will have much the characters of the deep-sea deposits, except in so far as we should expect to find it retain the special features, faunal and lithological, of shallow-water formations. These are said to be absent in the organic deposits immediately underlying the coral-limestone of the Solomon Islands, and it is accordingly concluded that they represent deep-sea formations. But the difficulty is not removed through this interpretation, since even if they are deep-sea deposits their elevation into the upper zone would have brought them within the reach of surface conditions. And yet the accompaniments of these conditions seem to be wanting until we reach the corals themselves. The negative character, therefore, gives no evidence as to the depth at which the sub-coralline deposits were laid down. But the fact that no coral rock is found at any really great elevation above the sea is sufficient evidence, it seems to me, that there has been no such marked elevation as

Mr. Guppy suggests, otherwise it would be almost impossible to account for the nearly equal altitude (above the water) which this formation holds in the different islands of the island group. Indeed, the fact that by far the greater number of coral-islands and reefs lie practically at the level of the sea, or but a few hundred feet above it in the case of fringing-reefs, is an almost insuperable objection to the theory which holds to the formation of atolls through elevation; the uniform line of position is opposed to any law of chances which might be assumed to govern a broad elevation. The same objection naturally does not apply to a theory of upgrowth in a stable area any more than it does in the case of a subsiding one.

Mr. Guppy has, indeed, himself anticipated some of the objections to his own views, but it appears to me he has failed to grasp their full significance. If, as it is claimed by the author, reef-building corals may thrive at a depth of 40, 50, or 60 fathoms, and if their structures are planted directly upon deep-sea deposits, then manifestly the thickness of the coral-made rock should be very much greater than has actually been found to be the case.

Mr. Guppy attempts to meet this difficulty by assuming [the immediately reversed position] that reef-corals will be usually confined to depths of less than 20 or 30 fathoms, and that the "rapid sub-aerial denudation, to which these regions of heavy rainfall are subjected, would be an important agency in the thinning away of the raised coral formations" (Proc. Royal Soc. Edinburgh, 1885–86, p. 890). This is surely begging the question—indeed, it might be said, it is abandoning the main proposition—since in the feeble development of the coral-made rock the one vulnerable argument against the Darwinian hypothesis was supposed to lie; it is in this fact that the opponents of the subsidence theory have intrenched themselves. Yet we have here the testimony of the only investigator in the premises that the thinness of the rock in question is probably not as thin as it is supposed to be; indeed, for any evidence that has been brought forward to the contrary, the rock may

have been very thick. In his review of the question of great
elevation Mr. Guppy thus expresses himself: "So great has
been the sub-aerial denudation of these islands, that, although
the elevatory movements have brought up to our view deep-sea
deposits which have been formed in depths probably of from
1000 to 2000 fathoms, yet, notwithstanding this great upheaval,
the calcareous envelopes, or ancient reef-formations, usually
disappear from the slopes of the large islands at heights of 500
or 600 feet above the sea, and never came under my observation at elevations much over 900 feet. . . . .
"Besides the testimony afforded by the stripping off of the calcareous envelopes from the higher levels, abundant evidence
of the great degradation which these islands have experienced
is to be found in the exposure at the surface in various
islands of highly crystalline and other much altered igneous
masses (such as quartz-diorites, quartz-porphyries, gabbros,
felspar-rocks, altered dolerites, and serpentines), which, according to Professor Judd and Mr. Davies, were formed and also
altered at great depths, and could only have been exposed by
extensive denudation. Of the rapid degradation of the surface
which the calcareous districts undergo in this region of heavy
rainfall, there can be no doubt. It should therefore be remembered, when examining this region, that although in post-
Tertiary times it has been an area of great upheaval, which a
moderate computation would place at not less than 12,000 feet,
it has also been an area of most rapid denudation" ("Solomon
Islands," pp. 125–26). After this admission of enormous waste,
the argument from the thinness of the coral-limestone loses all
force; nor can it be reasonably claimed that the waste extended
only over a thin and upwardly-extended capping of rock, since
Mr. Guppy assumes for one of his important conclusions that
barrier and other reefs grow out on their own talus. With origination in a great depth there would be ample opportunity for
such outward growth, and the accumulation of vast thicknesses
of rock. And how would rock accumulated in this way differ
from rock accumulated through subsidence? And if great

thicknesses of coral-made rock, whether formed in the one way or the other, have been removed from the elevated reefs of the Solomon Islands, wherein lies the evidence that there has been no subsidence?

H. B. Guppy. " The Solomon Islands." 1887.

"The Coral-Reefs of the Solomon Islands." Nature, Nov. 25, 1886.

"Observations on the Recent Calcareous Formations of the Solomon Group made during 1882–84."

The principal facts contained in these papers bearing upon the coral-reef problem have been considered in the preceding review of Mr. Guppy's paper "Notes on the Characters and Mode of Formation of the Coral Reefs of the Solomon Islands."

W. J. L. Wharton. " Masámarhu Island." Nature, Sep. 1, 1888.

A delineation of two slopes of the coral reef surrounding the small island of Masámarhu (situated in the Red Sea, in Lat. 18° 49′ N. and Long. 38° 45′ E.), as determined by Captain Maclear, of H. M. S. "Flying Fish." This is an important contribution to the history of reef-structures, since it places beyond doubt the fact that the seaward slope of some coral islands is very abrupt, as earlier determinations had reported. At one point removed about 375 feet from the growing edge of the reef soundings indicated a depth of 1200 feet, or an average descent for this portion of the slope of some 72°. At a distance of 1200 feet the depth was found to be 1300 feet. Beyond this point the seaward slope is somewhat less abrupt, and at a distance of about 1900 feet a depth of only 1500 feet was found. Even this is a steep slope, averaging 38°, and fully equal to the slope of the steeper volcanic cones; the first portion of the descent, on the other hand, far exceeds the slope of any mountain-peak with which we are acquainted, except where sheer (so-called " vertical ") rock-precipices are presented. Coral and coral sand were obtained from nearly all parts of the slope, and at one point coral limestone was struck at 1300 feet. In two or three places the line dropped into deep and narrow ditches, the walls of which on both the inner and outer sides were very abrupt, rising at an inclination of about 80.° One of these ditches, reaching in its bottom to 1200 feet, has a depth, measured by the height of the outer wall, of upwards of 350 feet.

The facts of this island, so far as they go, are distinctly in favor of the subsidence theory, and they have been properly estimated by Prof. Bonney (Nature, May 23, 1889). Mr. Guppy, on the other hand (Nature, May 16, 1889), sees in Captain Maclear's sections evidences favoring Murray's views! How they favor these views is not stated, nor do I believe that it would be easy to find any confirmation in them of the theory of organic upgrowth. Of course it can be assumed that outward growth on an extended talus *might* (under special conditions) produce such a steep slope, but this is far from proving that the condition did in fact exist. Further, we should still be compelled to prove that any such large talus can form (and I believe Prof. Dana has well argued that it *cannot* readily form), and that even if formed, there is that (vast) outward growth upon it which has been assumed by Murray and Guppy. As regards his own special views of the formation of barrier-reefs, etc., Mr. Guppy finds full confirmation in the "ditches" which were located on the slopes of Masámarhu, for he thus expresses himself: "The 'ditches' shown in these sections I look upon as indicating the formation of barrier-reefs at considerable depths, and as giving remarkable support to my views on the origin of these reefs" (Nature, May 16, 1889). Mr. Guppy has, indeed, pointed out (in a very unsatisfactory manner) that reef-building corals *may* thrive at depths of 50 or 60 fathoms (300–360 feet), and that barrier-reefs and atolls *may* begin to build up from these depths (answering the difficulty with regard to the deep lagoons and channels); but now we are suddenly called upon to assume that they build up from of a depth of 200 fathoms (1200 feet)! Surely the most doubtful cannot readily object to a theory which is so elastic as that of coral upgrowth.

H. B. Guppy. " Preliminary Note on Keeling Atoll, known also as the Cocos Islands." Nature, Jan. 3, 1889.

In this paper (letter addressed to Mr. Murray) the author presents some interesting facts pertaining to the formation of horse-shoe shaped atollons. His conclusion may be briefly stated : " that wherever a coral island stems a constant surface-current, the sand produced by the breakers on the outer edge of the reef will mostly be deposited by the current on each side of the island in the form of two lateral banks or extensions, giving the island ultimately a horse-shoe form, with the convexity presented against the current."

A bank may then be " thrown up across the mouth of the horse-shoe, and a small atoll with a shallow lagoonlet is produced." Other points reached in Mr. Guppy's examination of the Keeling atoll are that " the lagoon is rapidly filling up with sand and coral " and " that the outward extension of the reef is effected, not so much by the seaward growth of the present edge of the reef, as by the formation outside of it of a line of growing corals, which, when it reaches the surface reclaims, so to speak, the space inside it, which is soon filled up with sand and reef-debris."

Mr. Guppy prefaces his paper with an appeal to the shallowness of lagoons, and criticizes our exaggerated notions of these structures. On a true scale it is claimed that a typical lagoon " would be represented by a film of water occupying a slight hollow in the level mountain-top." The author further expresses himself as follows : " By thus grasping these facts, we at once perceive that by reason of our failing to view an atoll in relation to its surroundings, and through our misconceptions of its dimensions, we have been led to introduce a great cause to explain a very small effect. The slightly raised margins can be easily explained by causes dwelt upon by Murray, Agassiz, and others. No movement of the earth's crust is necessary for this purpose. The mode of growth of corals, the action of the waves, and the influence of the currents, afford agencies quite sufficient to produce the slightly raised margins

of an atoll." It is, indeed, hard to class the logic of this argument. Is it to be presumed that because some geologists have
exaggerated notions of the configuration of an atoll and its inclosed lagoon, that the lagoon does *not* exist? And if it really
does exist, how near are we brought to an understanding of
its structure by the simple conception of its being a thin film
of water perched upon the summit of a flat mountain-top?
The depth of the lagoon still remains the same, and so does
the height of the raised border. It might, indeed, as well be
urged that there is no necessity to account for the structure of
mountain-chains, since the highest of them are mere wrinkles
on the earth's surface, corresponding in size to the irregularities on the rind of an orange! Comparisons are useful, but
they are not explanations.

Johannes Walther.   "Die Korallenriffe der Sinaihalbinsel." Abhandl. d. mathem.
     physisch. Classe der Königl. Sächs. Gesellsch. der Wissenschaften, xiv, 1888.

No new facts tending toward the solution of the coral-reef
problem are given in this valuable memoir. The coral-structures described are mainly fringing-reefs, and they occur in a
region of existing or recent elevation. The author calls attention to certain atoll-like or ring-formed islets which are found
associated with the fringing-reefs, and which in some instances
are immediate outgrowths from the latter. This circumstance
is immediately seized by Mr. Bourne (? G. C. B.—Review of
Walther's work in Nature, Dec. 20, 1888) as another instance
"added to the many now accumulating of barrier reefs and
atolls being formed in an area of elevation." Walther gives
no data regarding these islets, except as to form. It seems
proper to ask in this connection: Are all circular or crescen-
tic growths of coral to be classed as atolls? Are fringing-reefs
which for a short distance leave the coast-line to be classed as
barrier-reefs? Are the dunes and sand-hills of the continental
areas mountains?

As touching the question of the formation of these islets in
an "area of elevation" it is interesting to note that Walther
recognizes a local subsidence of some 6 metres at the southern

end of the Peninsula (Rás Muhámmed), *or almost in the very region of the islets in question.* Further to the north, on the other hand, the reefs are strictly linear and conform rigidly to the coast-line, and are confined within the 10-fathom line. In the region of the islets, opposite to the points where subsidence is supposed to have taken place, the depths are much greater.

S. J. Hickson. "Theories of Coral Reefs and Atolls." Address British Assoc., 1888.

Some points contained in this paper, which is a broad review of the opposing theories of reef-formation, have already been noticed. Prof. Hickson considers himself an adherent of the views of Mr. Murray, but feels doubtful about two points: (1) "Whether the power of solution of sea-water is sufficient to account for the formation of lagoons, and (2), whether in some cases, such as the eastern part of the Feejee Archipelago and the Low Archipelago, the theory of subsidence may not be the correct one." The author believes that from the evidence of the Great Chagos bank alone "the subsidence theory breaks down," but he gives no facts to support this position beyond the belief that the banks are (or ought to be) rising instead of subsiding, as they were considered to be by Darwin. The all-important fact which Darwin pointed out, on the testimony of Captain Moresby, that the rim of this supposed "drowned atoll," lying at a depth of a few fathoms beneath the surface of the water, consisted almost wholly only of dead coral, Prof. Hickson believes "requires re-investigation," because "it is difficult upon any theory to see why the rim only nine or ten fathoms below the surface should not be covered with live coral." I fail to see why this assumed fact is opposed to "any theory" of coral growth, except the one which has been advanced or sustained by Murray, Guppy, and Bourne. While it is admitted by both Darwin and Dana that the reef-building zone extends down to a depth of 20–25 fathoms, it is well known that in many coral regions the practical limit of coral development is found at only half this depth. Thus, during the cruise

of the Wilkes Exploring Expedition the anchor of the "Pea-
cock" was dropped [within the reefs of Viti Lebu and Vanua
Lebu] sixty times in water from twelve to twenty-four fathoms deep, and in no case struck among growing corals.
Patches of reef were encountered at times, but they were at a
less depth than twelve fathoms" (Dana, "Corals and Coral Islands, 1872, p. 116). Alexander Agassiz's researches off the
Tortugas reefs lead to the conclusion "that corals do not thrive
below a depth of from six to seven fathoms," and the same
limit of growth was found by Louis Agassiz along the whole of
the main reef to the northward" ("Three Cruises of the Blake,"
vol. I, p. 74, 1888). Walther found dead reef (supposed to have
been brought about by subsidence) at Râs Muhámmed at a
depth of only 10 metres, or 30 feet (" *Die Korallenriffe der Sinaihalbinsel*," p. 465, 1888). Why then should the observations of
such a careful investigator as Captain Moresby be called into
question? It seems to me that they very closely agree with
the theory of subsidence.

H. O. Forbes. " A Naturalist's Wanderings in the Eastern Archipelago," 1885.

The author describes the Keeling atoll, which he believes
to have risen through elevation, and finds reason to conclude
that the lagoons have filled in materially since the time of Darwin's visit.

J. D. Dana. " Origin of Coral Reefs and Islands." Am. Journ. Science, 3d ser., vol. xxx,
1885.

An elaborate review of the objections raised to the Darwinian theory, and a statement of facts in support of said theory.
This is by far the most searching analysis of the divergent
views bearing upon the theories of coral-reef formation, and,
as it appears to me, it satisfactorily meets all the objections
that have been raised against subsidence, besides showing the
inadequacy of the substitute theory. The author's main conclusions are thus stated (p. 190): "The subsidence which the
Darwinian theory requires has not been opposed by the mention of any fact at variance with it, nor by setting aside Dar-

win's arguments in its favor; and it has found new support in
the facts from the Challenger's soundings of Tahiti that had
been put in array against it, and strong corroboration in the
facts from the West Indies." The main points contained in
this paper are such as have already been considered, and require no lengthy discussion in this place. Prof. Dana is emphatic in his belief that subsidence (preceded by elevation)
was the condition which permitted "of the making of the
Florida, Bahama and other West India coral reefs," a view in
which he is distinctly opposed to Mr. Agassiz. The evidences
for this subsidence in a comparatively recent period are found
in the mammalian remains of apparently Quaternary age
which have been discovered in Cuba and Anguilla, and which,
from their special characters, point to a former connection
between these islands and the mainland. The belief in
a connection between the Windward Islands and the South
American continent has also been held by Cope and Pomel.
Dr. Supan, in reviewing Prof. Dana's paper (*Petermanns Mit-
teilungen*, vol. 32, pt. 1, *Litteraturbericht*, p. 5, 1886), criticizes
the views relative to subsidence in the Floridian region,
since, it is claimed, even if direct connection did exist between
the West Indian Islands and the southern continent, there is
no proof that this connection extended northward to the
North American continent; and he further denies—without,
however, giving any reasons for this denial—that there ever
was any (Quaternary?) connection between the West Indies
and North America. This notion is probably based upon the
old idea (advanced by L. Agassiz and Le Conte) of the making
of the Floridian peninsula, in which no movements of either
elevation or subsidence were supposed to have been involved.
Since, however, this conception has proved to be a myth there
is no further reason, except in so far as the case may be supported by fact, to adhere to the old views of continental (or
oceanic) stability in this region. My own observations have
conclusively proved a peninsular uplift as late as the Post-
Pliocene period, and extending as far south as Lake Okeecho-

bee. But I am by no means convinced, as I have elsewhere stated (chapter on the "Coral-Reef Problem"), that a nearly simultaneous subsidence did not take place in (and form) what are now known as the Straits of Florida. The existence of such a subsidence *Bruch* is considered likely by Suess (*Antlitz der Erde*, vol. I), who has paralleled it with (supposed) similar occurrences in the eastern basin of the Mediterranean. This view of the formation of the deep Gulf-channel, I must confess, appears to me far more captivating than that which ascribes it to the wash of the Gulf-current.

But I believe direct evidence pointing to (although by no means proving) a former connection between the Floridian peninsula and the mainland to the south is not wanting. In a paper on "The Value of the 'Nearctic' as one of the Primary Zoological Regions," published in the Proceedings of the Academy of Natural Sciences of Phila. for 1882, I pointed out certain facts in favor of considering the lower portion of the peninsula as a part of the Neotropical, rather than of the Nearctic, realm; more recent zoological researches have still further demonstrated the correspondence existing between this southern fauna and that of the tract lying to the south. But more significant is the finding of the large assemblage of mammalian remains which have lately been brought to light from various parts of the peninsula. These have been determined by Dr. Leidy to be the skeletal parts of the elephant, mastodon, llama, rhinoceros, tapir, Hippotherium, the sabretooth tiger (Machairodus), Glyptodon, etc. (Leidy: Proc. Acad. Nat. Sciences of Phila., 1884–89). Neither the sabre-tooth nor the Glyptodon, both of which are so closely related to the commoner South American forms as to be barely distinguishable from them, have heretofore been found in the Southern United States. Of course they may yet be found, and indicate a passage over from South America by way of Mexico and the Southern United States. But the great abundance of these remains on the Floridian peninsula, and their absence either in whole or in part from the Gulf States, are facts which, so far as they go, point to a former direct land-connection across what is now an arm of the Gulf.

J. D. Dana. "Points in the Geological History of the Islands of Maui and Oahu." Am. Journ. Science, 3d ser., xxxvii, 1889.

The author gives the results of artesian borings made on Oahu (Sandwich Islands) which indicate the presence of coral-rock at depths varying from 500 feet or less to upwards of 1000 feet beneath the level of the sea. In Mr. Campbell's well, west foot of Diamond Head, a continuous bed of coral, 505 feet in thickness, was struck at a depth of 320 feet. The species of coral found in these deep rocks not having been determined, Prof. Dana holds that some doubt may yet be entertained as to the beds in question affording positive proofs of subsidence, although there is a strong probability favoring this view. It is interesting to note in this connection that in the deep well above noted a soapstone-like rock, 20 feet in thickness, was found immediately underlying the basal bed of coral at a depth of 1048 feet. Is this the correspondent of the soapstone-like beds which Mr. Guppy found at many points underlying the coral limestone of the Solomon Islands, and which that investigator considered to be evidence in favor of the view that corals began to grow upward from great depths? This point is fully discussed in the review of Mr. Guppy's papers.

R. von Lendenfeld. A review of Mr. Bourne's paper on the Diego Garcia Reef. Naturwissenschaftliche Rundschau, Oct. 13, 1888.

The author finds no facts either in this paper, or in the papers of Murray, Agassiz, and Guppy, which are opposed to the theory of subsidence. He justly calls attention to the low-level of coral islands generally, which is opposed to any theory of the progressive elevation of the ocean bottom. The great development of the dolomite reefs of southern Tyrol (of the Rhætic period), such as the Langkofel, are cited in evidence against the assumed non-existence of thick coral limestones in any of the older geological formations.

It is important to note in this connection the observation by Suess that nowhere in the Gosau deposits, nor in the coraliferous beds of Cormons (Eocene) or Crosara and Castel Gomberto (Oligocene), did he observe any evidence of the existence of massive coral structures, such as might be strictly compared to the modern reefs. (Antlitz der Erde, ii, p. 407, 1888.)

A. Geikie. Presidential Address delivered before the Royal Physical Society of Edinburgh. Proc. Royal Phys. Soc. Edinburgh, viii, 1883-84.

A review of the rival theories of coral-reef formation. Prof. Geikie gives in his adhesion to the views of Mr. Murray, but adduces no fact which is inconsistent with the theory of subsidence.

J. Rein. "Die Bermudas-Inseln und ihre Korallenriffe, nebst einem Nachtrage gegen die Darwinische Senkungstheorie." Verhandl. d. erst. deutsch. Geographentags, 1881.

I have not seen this paper, but the facts contained in it, so far as the Bermudas are concerned, seem to be the same as those which are given in his original memoir, and which are discussed in the main part of this work.

K. Semper. "Animal Life as affected by the Natural Conditions of Existence." 1881.

In this work the author reviews his observations on the coral-structures of the Pelew Archipelago—to which reference is made in the chapter on the "Coral-Reef Problem"—and asserts his positive conviction that the evidence obtained at this point is directly opposed to the theory of subsidence. The principal basis for Semper's conclusions is that we have here an association of all three classes of reefs—atolls, fringing-reefs and barrier-reefs. These he considers to have arisen in a region of elevation. Semper has, I believe, given us positive proof of elevation, but I fail to see how this elevation in any way precludes the possibility of a subsequent subsidence; nor can I find any facts in the description of the islands which speak against such a subsidence. On the contrary, the author himself admits that a change or arrest of movement (of elevation) has taken place when he asserts: "The facts here adduced suffice, as it seems to me, to prove that, in the first place, a quite recent upheaval must have occurred; and secondly, that that period of upheaval must have passed into the present con-

dition of very slow elevation or absolute rest without any conspicuous break" (p. 263).   If we assume that this arrest of
movement, passing into a condition of *absolute rest*, had still
further progressed into one of subsidence, I think we will then
be able to understand all the special (and apparently antagonistic) features of the region to which Prof. Semper refers, and
in a manner much more satisfactory than is offered by the explanation of the distinguished German naturalist.   It seems
to me that the condition here is almost precisely what we find
in the Bermudas: a coral-made land, which had been elevated
to some little height above the sea, undergoing waste and destruction through subsidence.   This phase in the history of
Bermuda is so clear that there can be no question concerning
it.   In the Bermudas we have also a near-lying reef on the one
side (likewise the weather side) and a far-off reef on the other,
with an intermediate body of water of some 50–60 feet depth.
But Prof. Semper himself gives data which lead one to suppose that subsidence has in fact taken place.   The biting out
or undercutting of the limestone plateau to which he calls attention (p. 254), and which may be paralleled with the similar
process on the south shore of the Bermudas, surely argues
much more strongly in favor of subsidence than of elevation;
it certainly seems impossible for a rock to be at the same
time building up and breaking down.   But further, Prof.
Semper informs us that it "is certain that the enclosed island
of Babelthuap was formerly much broader than it now is"
(p. 270). This condition is scarcely compatible with any theory
of elevation, and more nearly accords with the assumption of
waste through oceanic encroaches permitted by subsidence.
Indeed, the author himself seems not to have been convinced
that there was no subsidence, since he asserts that "we are
obliged, under all the circumstances, to assume the co-opera-
tion of some other force besides subsidence when endeavoring
to explain the peculiar formation of the northern reef, but still
without wholly excluding the effects of subsidence" (p. 253).

Prof. Semper rejects the hypothesis that the lagoon-ring of
an atoll is formed through accelerated marginal growth (p.
227); he attributes the lagoon-basin to decay and scour.

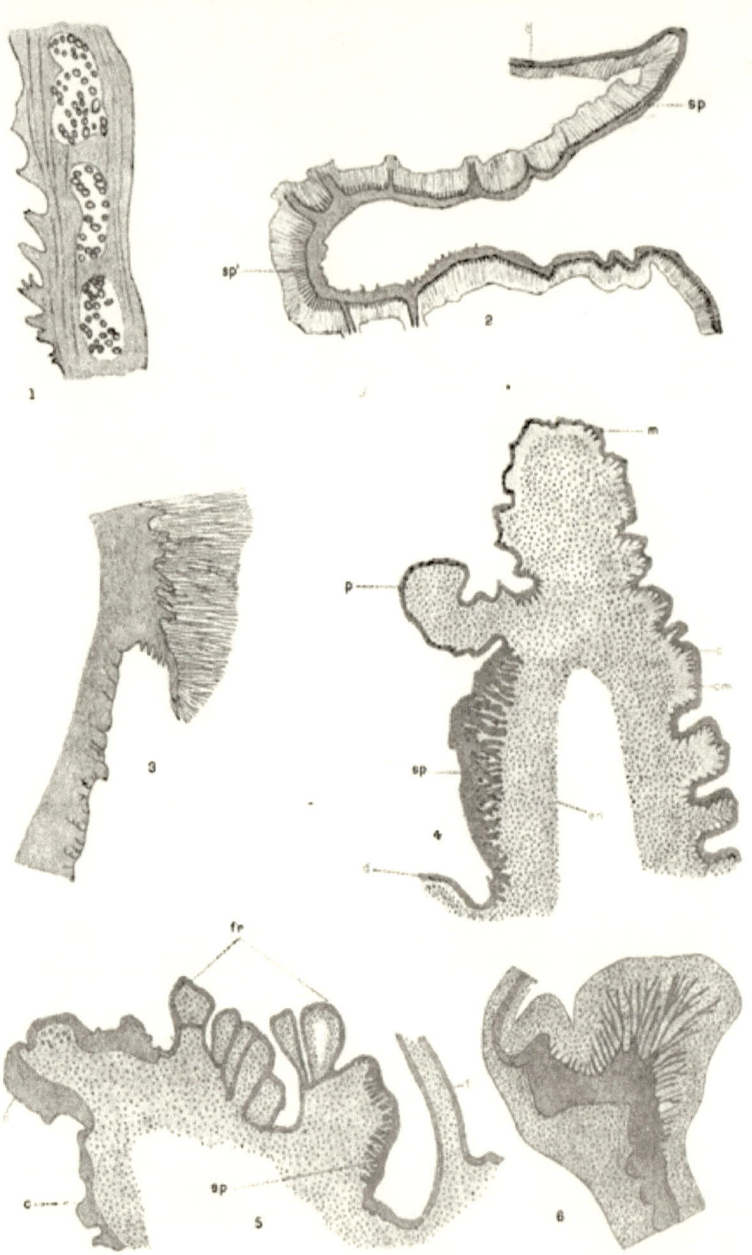

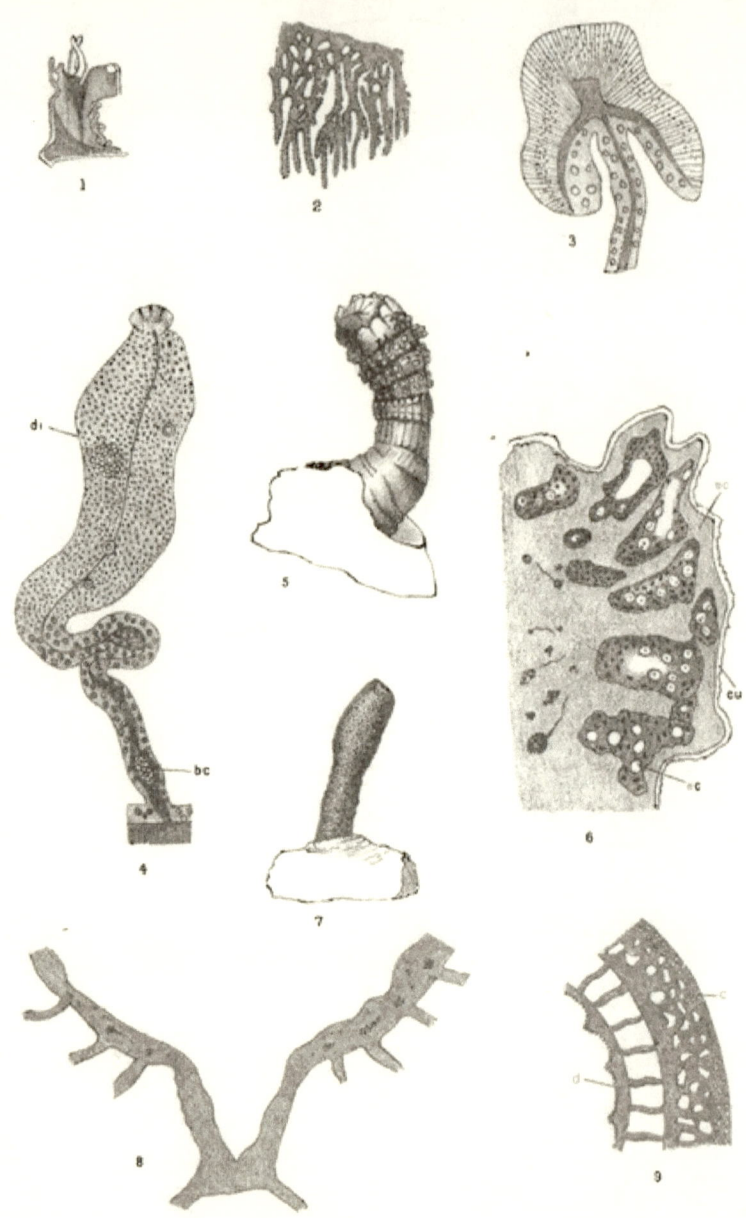

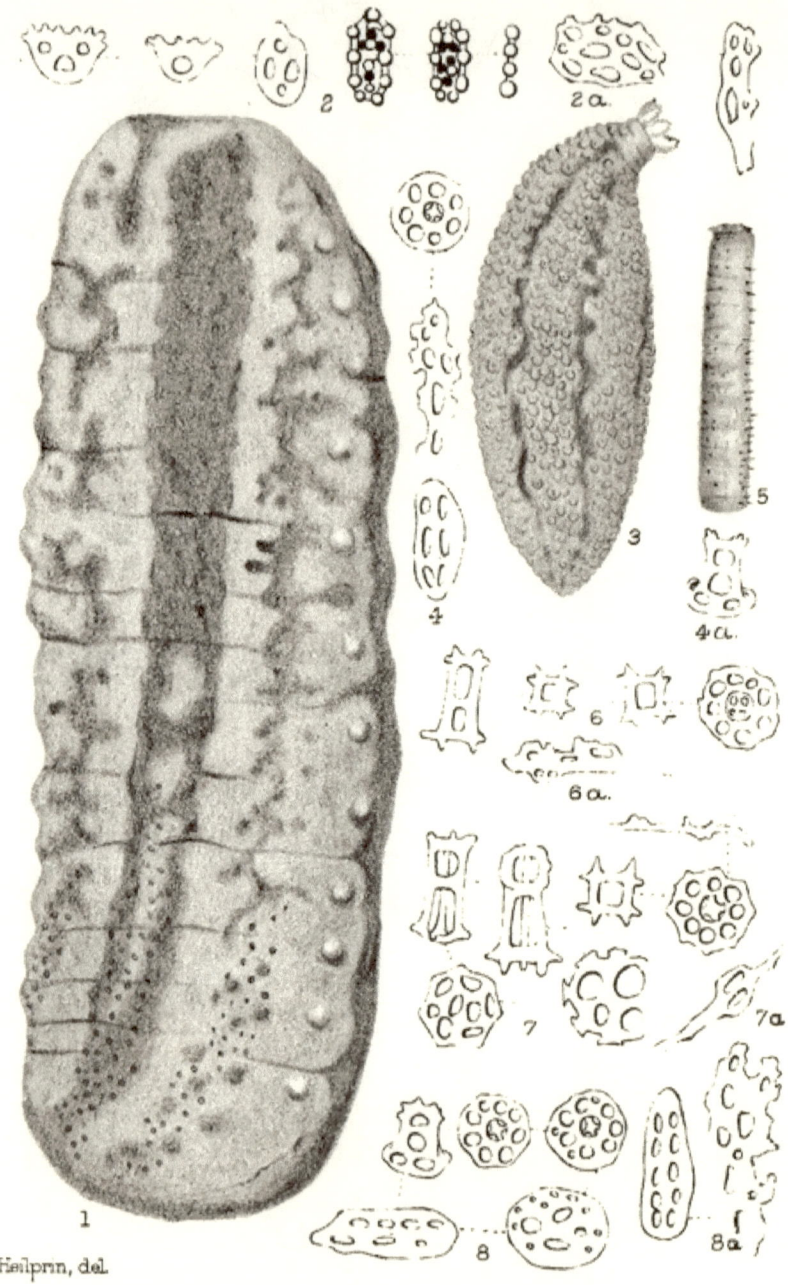

Pl. 12

Heilprin, del.

Zoology of the Bermudas.

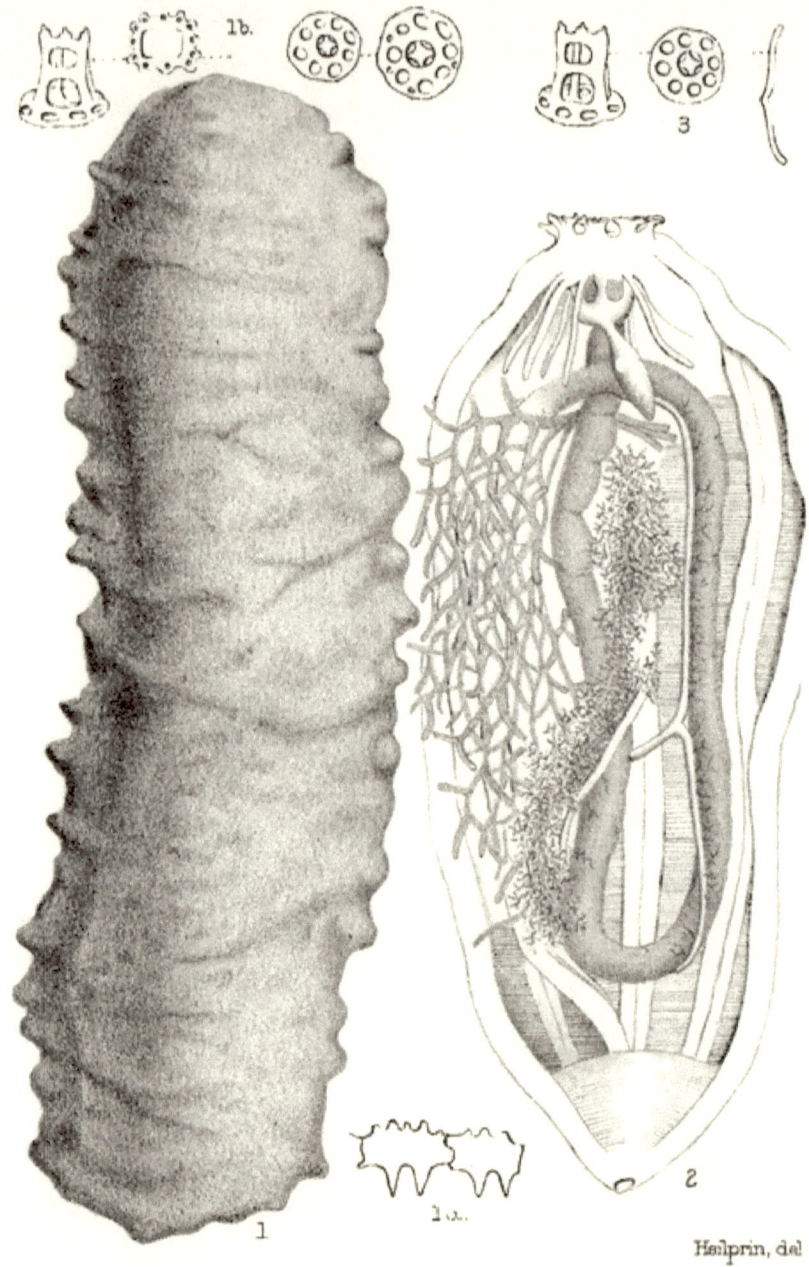

1b.

3

1

1 a.

2

Heilprin, del

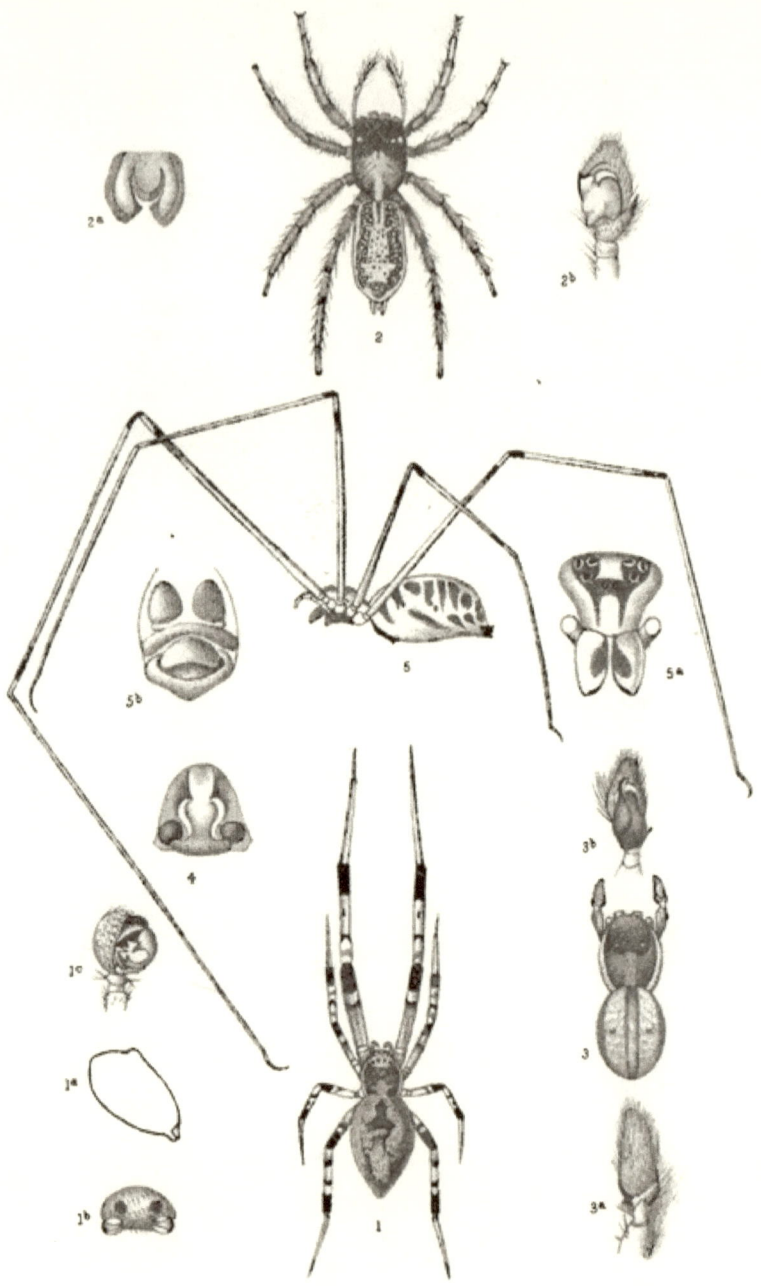

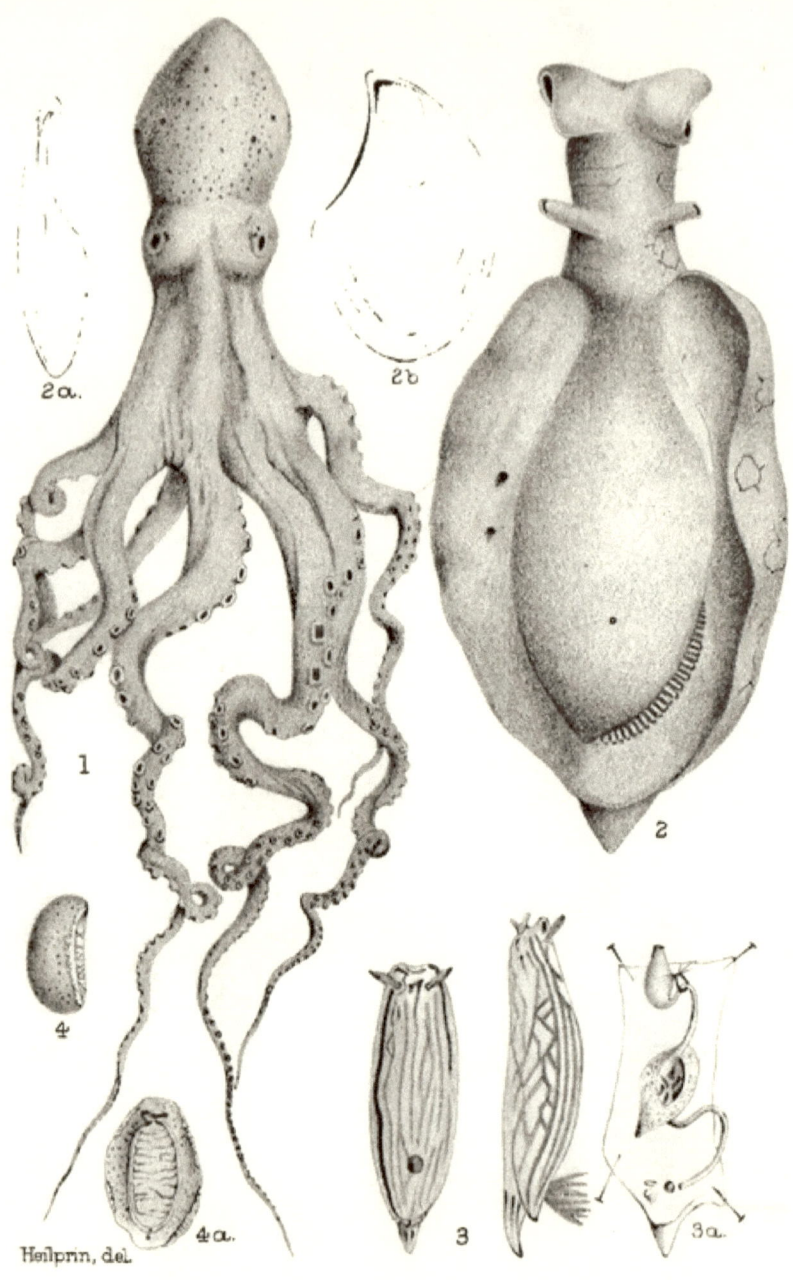

2a. 2b.

1

2

4

4a.

3 3a.

Heilprin, del.

Zoology of the Bermudas.

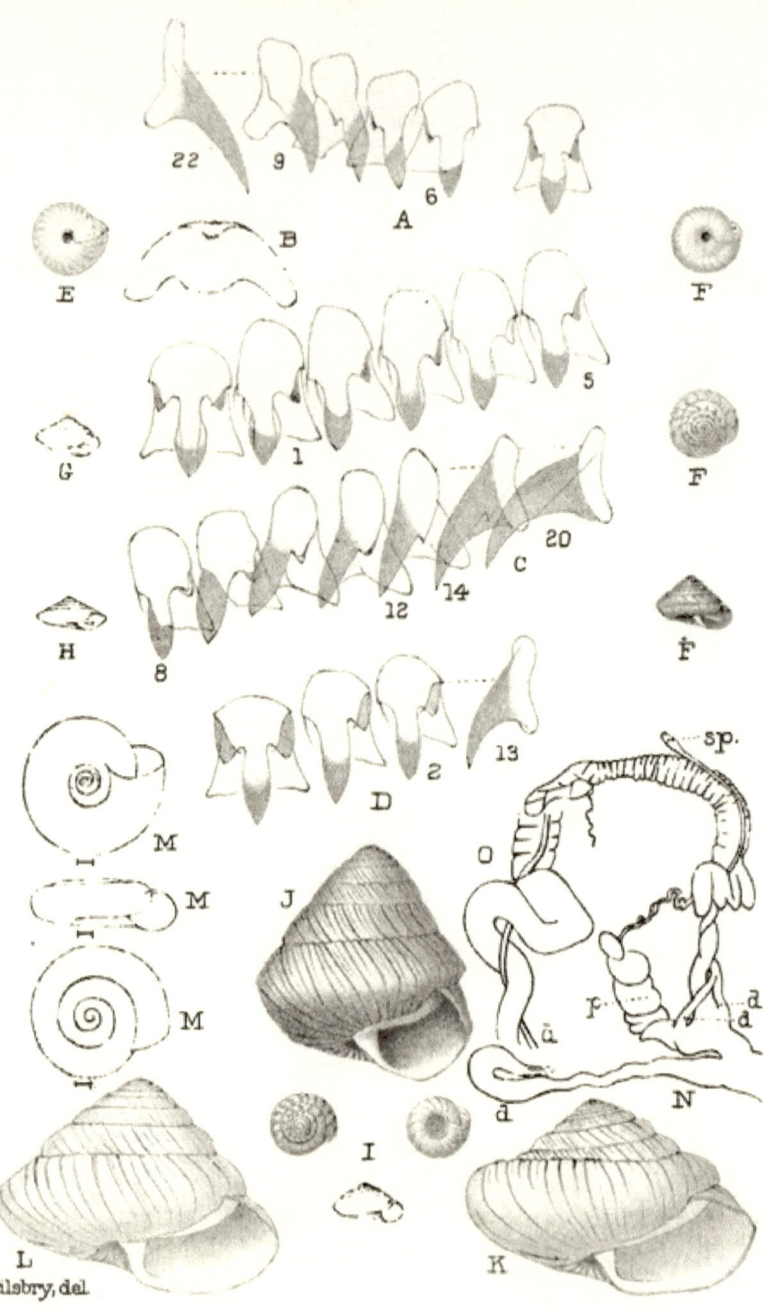

Zoology of the Bermudas.

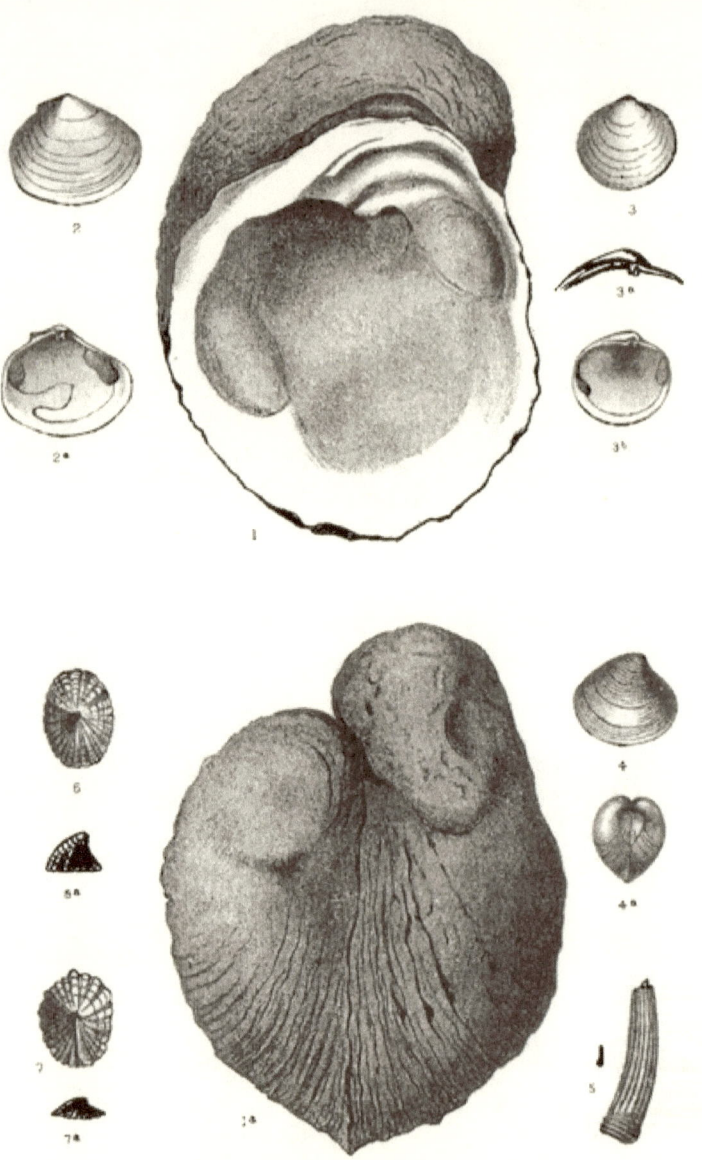